CAMBRIDGE STUDIES IN ECOLOGY

Editors:
R. S. K. Barnes *Department of Zoology, University of Cambridge*
H. J. B. Birks *Botanical Institute, University of Bergen*
E. F. Connor *Department of Environmental Science, University of Virginia*
J. L. Harper *School of Plant Biology, University College of North Wales*
R. T. Paine *Department of Zoology, University of Washington, Seattle*

Diseases and plant population biology

Also in the series

Hugh G. Gaugh, Jr. *Multivariate analysis in community ecology*

Robert Henry Peters *The ecological implications of body size*

C. S. Reynolds *The ecology of freshwater phytoplankton*

K. A. Kershaw *Physiological ecology of lichens*

Robert P. McIntosh *The background of ecology: concept and theory*

Andrew J. Beattie *The evolutionary ecology of ant–plant mutualisms*

Diseases and plant population biology

JEREMY J. BURDON

Division of Plant Industry, CSIRO, Canberra, Australia

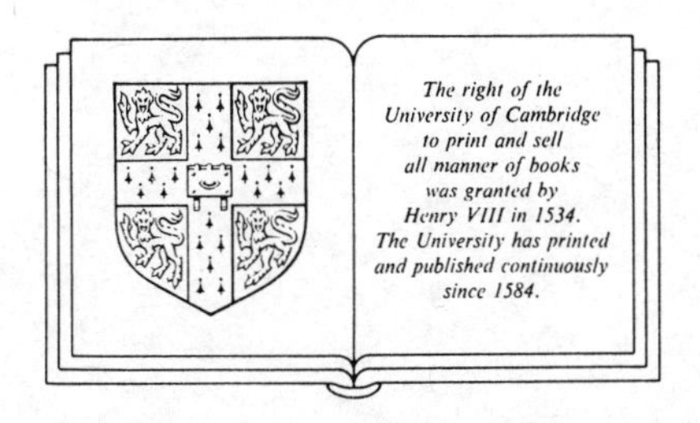

CAMBRIDGE UNIVERSITY PRESS

Cambridge

London New York New Rochelle

Melbourne Sydney

Published by the Press Syndicate of the University of Cambridge
The Pitt Building, Trumpington Street, Cambridge CB2 1RP
32 East 57th Street, New York, NY 10022, USA
10 Stamford Road, Oakleigh, Melbourne 3166, Australia

© Cambridge University Press 1987

First published 1987

Printed in Great Britain by the University Press, Cambridge

British Library cataloging in publication data

Burdon, Jeremy J.

Diseases and plant population biology – (Cambridge studies in ecology)

1. Plant populations 2. Micro-organisms, Pathogenic

I. Title

581.5′248 QK911

Library of Congress cataloging in publication data

Burdon, J. J. (Jeremy James)

Diseases and plant population biology.

(Cambridge studies in ecology)

Bibliography

Includes index.

1. Plant diseases. 2. Plant populations.

3. Micro-organisms, Phytopathogenic – Host plants.

I. Title. II. Series.

SB731.B79 1987 581.5′248 86-19269

ISBN 0 521 30283 8 hard covers
ISBN 0 521 31615 4 paperback

UP

Contents

Preface

Population biologists have strangely discordant views concerning the part herbivores and pathogens play in plant communities. The role of animal grazers in affecting the dynamics and composition of individual communities is well accepted. In agricultural situations, careful management of stocking rates has often been used to modify the botanical composition of pasture communities, while in less controlled systems a burgeoning body of literature attests to the diverse ways in which phytophagous insects (surely the zoological equivalent of plant pathogens) may affect the fecundity and survival of individual plants.

By contrast, interest in the role of plant pathogens (regarded here as including bacteria, fungi and viruses) is characterized by its agricultural bias. Although plant pathogens have been the immediate cause of many of the more spectacular crop failures of history, they are not uniquely associated with intensive farming practices. Diseases caused by a range of pathogens are to be found in most plant communities. Despite this, however, plant biologists have generally ignored the ways in which pathogens affect the size and genetic structure of individual plant populations and, through this, the composition of entire communities. In writing this book I hope to begin to rectify this imbalance.

Studies involving interactions occurring between two such diverse groups of organisms as plants and their pathogens are, of necessity, multi-faceted. Here I have tried to walk the interdisciplinary tightrope between ecology, genetics and plant pathology to provide a general guide to the many fascinating ways in which plant populations may be affected by pathogens. In doing this I have also tried to provide some insight into the relevance of host–pathogen interactions for diversity theory, for considerations of the maintenance of polymorphisms in plant populations, and for the occurrence of co-evolutionary interactions in nature. Primarily, I have had non-agricultural populations in mind and, ideally, would have liked to include only examples taken from studies in natural systems. This was entirely impractical however, although I have made every effort to use such examples wherever possible.

This book has been written specifically for population biologists, both ecologists and geneticists. For this reason I have deliberately avoided covering much of the general information concerning the population biology of plants that has been reviewed so well by Harper (1977) and more recently by Silvertown (1982). Conversely though, much of the early part of the book deals with the basic 'nuts and bolts' of the effects of pathogens on individual host plants. However, as my primary interest here is in illustrating the effects of pathogens on the population biology of their hosts, I am sure that this general coverage of pathogen activity will not have provided sufficient detail to satisfy all plant pathologists. For this I can only apologise.

Over the last decade and a half my ideas concerning the role of pathogens in plant communities have benefited greatly from long discussions, arguments and associations with Professors J. L. Harper and the late I. A. Watson and Drs G. A. Chilvers, W. A. Heather, R. Whitbread, R. C. Shattock, R. H. Groves, D. R. Marshall, A. H. D. Brown, G. A. Kirby, N. H. Luig, R. A. McIntosh, H. M. Alexander, J. V. Groth and A. P. Roelfs.

I am particularly grateful to Greg Kirby, John Harper, Irvine Watson, Andrew Jarosz, Tony Brown and Richard Groves for reading, criticizing and improving sections of the manuscript and for providing much needed encouragement in dark days. However, all remaining faults are mine alone. Above all, I thank my wife Jill for providing continuing support without which this task would not have been finished.

March 1986 J.J.B.

1

Pathogens and the population biology of plants

While it has long been acknowledged that plant pathogens have had a major role in shaping the development of modern agriculture, there has been little recognition of the part they play in shaping non-agricultural plant communities. This book has been written with the intention of redressing, at least in part, this imbalance. Pathogens are found in most plant communities although their effects may be restricted to locations or seasons in which the environment is particularly favourable for their growth and development. Because they can be very damaging and are frequently highly selective, debilitating one individual while leaving adjacent plants untouched, plant pathogens are an important part of the biotic forces that shape plant communities.

By reducing competitive vigour or even killing individual plants, pathogens may affect the outcome of intra- and inter-specific competition, the distribution of plant species, the genetic structure of populations, the diversity of individual plant communities and possibly even the evolution of sex. A first step towards an understanding of the diverse roles played by pathogens in the ecology of plants requires a consideration of the basic demographic changes that occur in plant populations as they grow and mature.

Life and death in plant populations

In recent years plant ecologists have enthusiastically embraced the idea that plant populations are dynamic entities. The demographic changes that occur within individual populations have now been documented for a wide range of different species (Harper, 1977; Silvertown, 1982). The survivorship curves which typically embody the results of such recruitment studies provide a clear picture of the attrition that occurs as individual populations germinate, establish, grow and mature (Figure 1.1). However, these curves provide no indication of why given individuals

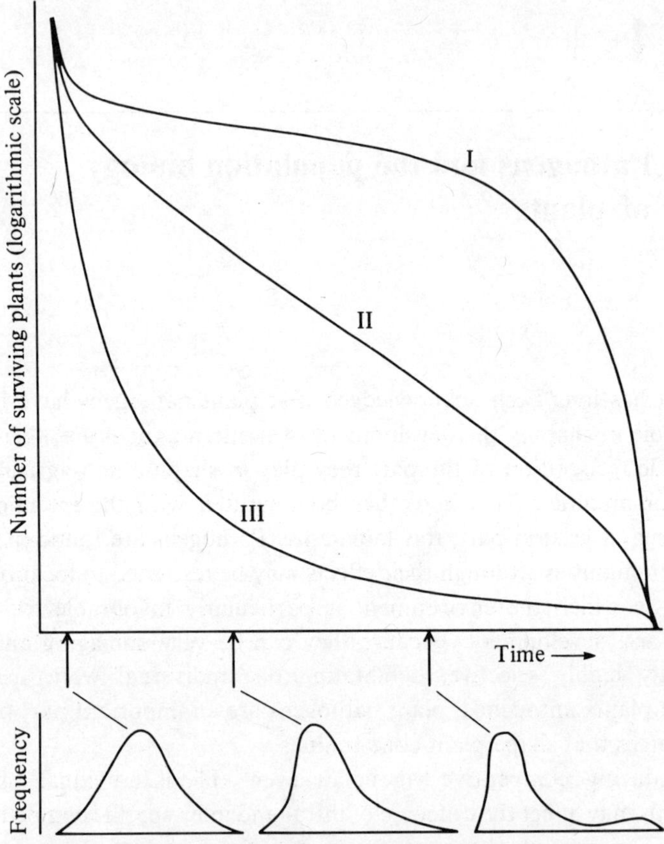

Figure 1.1 The size and structure of developing plant populations. (Top) Three typical but contrasting survivorship curves found among plant populations. These correspond to the classic Deevey curves I, II, and III (Deevey, 1947) and show how the timing of periods of maximum mortality may vary. (Below) The development of a size hierarchy in plant populations as they mature. Such hierarchies are likely to develop in all populations in which individual members compete for limited resources.

die. In general, patterns of death change markedly as populations develop. During the germination and establishment phase of growth, some individuals disappear without trace within the space of just a few days. Later, as the remaining plants come into closer and closer contact and competition intensifies, other individuals almost imperceptibly fall behind the growth of their fellows. These plants enter a twilight zone of increasingly suppressed individuals, some of which eventually disappear and others of which survive to reproduce, although their fecundity is usually much reduced.

The basic sequence of events that occurs in this development of a hierarchy of dominance and suppression is now well recognized. As a dense stand of plants grows, the size class distribution of individuals comprising the population becomes more and more skewed and relatively few individuals come to dominate the majority. Eventually a point is reached at which continued growth of the population as a whole is achieved only through the death of some individuals. Detailed studies have shown that this 'self-thinning' mortality response of pure species stands to highly competitive conditions is remarkably consistent between different plant species (Yoda *et al.*, 1963; White & Harper, 1970; White, 1980). While interference between adjacent individuals may start at an early stage, mortality itself becomes more noticeable when the relation between the logarithm of mean plant weight and the logarithm of stand density has a slope of $-3/2$.

These studies implicitly or explicitly emphasize that death, suppression and reduced fecundity are all features typically associated with the growth and development of plant populations, whether these are growing in the field or in carefully controlled glasshouse situations. This inevitably concentrates attention on the question of what factors determine the fate of individuals in a population. Why do some individuals live and others die? Furthermore, among the survivors why do some plants prosper while others struggle on the very edge of existence?

Despite the extensive documentation of demographic processes occurring in natural plant populations, studies that actually attribute deaths to specific causes (e.g. Mack & Pyke, 1984) are still exceptional. Few, if any, have attempted to document why particular individuals in such populations have become suppressed. The processes involved in the establishment of hierarchies of resource exploitation have been addressed in more controlled situations, but even there emphasis has been placed on inherent differences between members of the population. Minor variations in the time of emergence, seed size, seedling vigour and the relative growth rates of plants have all been shown to contribute materially to differences in the size of competing individuals (Black, 1957, 1958; Ross & Harper, 1972; Harper, 1977).

In the relatively unnatural environment of the glasshouse or carefully controlled field experiment such features are probably quite sufficient to explain the relative position of individual plants in a developing hierarchy and, hence, which individuals are subsequently culled from the population. However, plants do not live alone in the world. In reality they grow in environments teeming with other organisms, many of which regard plants as a food resource. By destroying buds, leaves, stems or roots these other organisms may totally upset established competitive relationships. Suppressed individuals may be released and dominant plants may be greatly reduced in size while yet other individuals may be totally destroyed.

Predators and pathogens are major biotic forces that can affect competitive interactions between plant species. As Harper (1977) has pointed out, it is an open question whether plant predators (including insects, molluscs, birds and mammals) are more or less important in this regard than pathogens. However, there is no doubt that the effects of predators on their host plants have caught the imagination of population biologists far more so than has the effect of pathogens. The role of plant predators has been the subject of numerous excellent studies (e.g. Tansley & Adamson, 1925; Jones, 1933; Waloff & Richards, 1977; Brown, 1982; Whittaker, 1982), reviewed in a wide variety of volumes (e.g. Rosenthal & Janzen, 1979; Crawley, 1983). In comparison, the effects of pathogens on plants growing in non-agricultural communities have been little studied.

The frequency of disease in natural plant populations

Perhaps one of the main reasons for the low degree of interest shown by ecologists in the role of pathogens in natural plant communities is their apparent lack of importance when compared with the occurrence of disease in agricultural systems. On occasions it has been suggested that disease epidemics are rare or virtually non-existent in natural plant communities while in agricultural systems they occur with monotonous regularity. This is a gross over-simplification. Differences in the occurrence and severity of disease between agricultural and non-agricultural systems are not absolute, rather they are differences of degree. Certainly, for much of the time pathogens are uncommon in wild plant populations but the same is also true for agricultural ones. Moreover, while disease outbreaks that cover large geographical areas and cause marked reductions in crop yields are particularly noticeable, such devastation is not a necessary pre-requisite for pathogen-induced changes in plant populations. In the long run, persistent low levels of disease may have a marked effect on the host population.

It is not true to claim that epidemics do not occur in non-agricultural plant communities. They do! The frequency of disease epidemics in natural systems seems to be correlated with both the relative antiquity and level of recent disturbance suffered by particular host–pathogen combinations. New combinations appear to be particularly prone to persistent, extensive disease outbreaks, as do host–pathogen associations occurring in ecosystems subject to recent high levels of disturbance. On the other hand, combinations that have developed over long periods of time tend to be characterized by epidemics of short duration and limited severity and spatial scale.

Most of the well-known, severe and geographically widespread epidemics that have been recorded in non-agricultural plant communities have occurred either in newly established host–pathogen combinations or in associations occurring in ecosystems subject to recent, marked disturbances. The destruction of North American stands of chestnut (*Castanea dentata*) by the pathogen *Endothia parasitica* (chestnut blight) and the visually obvious, yearly epidemics of *Puccinia lagenophorae* (rust) on *Senecio vulgaris* in Britain both result from the accidental introduction and establishment of pathogens into areas far outside their native range. Similarly, the radical alteration of existing ecosystems through clearing, selective felling and fire control has largely been responsible for producing conditions favourable for the development of persistent epidemics of *Cronartium fusiforme* (fusiform rust) on its native pine hosts in the south-eastern United States (Dinus, 1974).

Among the most widely recognized examples of periodic outbreaks of disease affecting long-standing host–pathogen combinations must be those occurring on weedy species occupying disturbed sites (e.g. *Albugo candida* (white rust) on *Capsella bursa-pastoris* in many parts of Europe and North America or *Puccinia hieracii* (rust) on *Taraxacum officinale* in Europe). However, epidemics of similar severity and duration also occur in many less disturbed communities. Browning (1974) reported that wild species of *Avena* growing in natural stands in Israel are, at times, heavily infected with *Puccinia coronata* (crown rust) and *P. graminis avenae* (stem rust). In Australia, outbreaks of rust (*Phakopsora pachyrhizi* and *Melampsora lini*) occur periodically on natural populations of *Kennedia rubicunda* and *Linum marginale*, respectively. Such observations are by no means unique. However, because these epidemics are limited, their occurrence and effect on host populations are generally underestimated.

It is the role that plant pathogens play in these sorts of situations that is the subject matter of this book.

The organization of this book

This book is concerned specifically with the role that fungal, bacterial and viral (including mycoplasmas and virus-like organisms) plant pathogens play in the ecology of their hosts. More is known about the role of pathogenic fungi and, consequently, it is this group which is mainly dealt with in this book. Where possible, examples involving bacteria or viruses are used to illustrate points which I believe are generally applicable to all plant pathogens. Other organisms that are often included in courses on plant pathology, for example nematodes, are not considered. Symbiotic relationships between bacteria or fungi and plants are, similarly, excluded from consideration.

It is difficult to understand the ways in which pathogens may affect plant populations without knowing something about their effects on individual plants. Chapter 2 examines some of these effects. During their life cycles, plants change dramatically in size and structural complexity. In doing so they provide a variety of different niches for potential pathogens. The various effects that pathogens may have as a result of attacking their hosts at different stages of the life cycle are discussed in the first part of Chapter 2. The second part of Chapter 2 addresses the major problem that is posed by the variable nature of pathogen-induced damage. The effects of the loss of a given amount of tissue on host survival and fecundity vary according to the host and pathogen involved, the part of the plant that is affected, and the duration and timing of the pathogen attack. Furthermore, even though disease may be restricted to specific sites its effects on the carbon balance and water requirements of the plant will often result in widespread changes in the metabolism and physiology of diseased plants. These complications highlight the difficulties encountered in trying to link disease severity to host fecundity.

Most plant pathogens have generation times that are considerably shorter than those of their hosts. During the course of a single season they may show enormous changes in population size. Chapter 3 considers the development of typical disease epidemics and the important differences which may occur between epidemics as a result of variations in the source of the initial inoculum and its mode of spread. The absolute size and rate of increase of a pathogen population is inevitably dependent on the host population and some of the ways by which host populations may affect those of the pathogen are explored in Chapter 3. This again emphasizes the importance of the source of inoculum on the patchy development of disease on plants growing in the wild.

Not all interactions between plants and their pathogens are restricted to numerical changes and in Chapter 4 the phenotypic expression and genotypic control of a wide range of specific and non-specific resistance mechanisms are reviewed. While relatively little is known about the genetic basis of non-specific resistance, that of specific resistance (and its corollary in the pathogen – virulence) has been extensively investigated and provides the genetic basis for most theories of co-evolution between plants and their pathogens. This theme is elaborated in Chapter 7. In the latter part of Chapter 4 the way in which this knowledge can be used to make a minimum estimate of the diversity present in host and pathogen populations is discussed.

The interactions occurring between host plants and their pathogens are not determined solely by the genetic constitution of the conflicting parties. The occurrence, expression and severity of disease symptoms are all affected to some degree by variations in the physical environment, the nutritional status and ontogenetic age of potential hosts, and by the presence or absence of a wide array of other organisms. These potentially confounding influences are referred to in all chapters but are considered in detail in Chapter 5.

Chapters 6, 7 and 8 consider the changes that may be brought about in plant and pathogen populations as a result of their interactions. The expectations of theoretical models and the empirical knowledge currently available concerning the effects of pathogens on the size of host populations is explored in Chapter 6. This leads into a consideration of the consequences of their differential actions in mixed plant populations and the role pathogens may play in determining micro-geographical patterns of distribution of individual species. A similar arrangement in Chapter 7 concerns changes in the genetic structure of host populations. This repeatedly demonstrates that the extent to which such changes are likely to occur depends primarily upon the suitability of the environment for the development of the pathogen population. However, the breeding system of the host plant involved appears to have an important influence on the extent and speed with which plant populations respond to pathogen-induced selective pressures.

This book is primarily about the effects of pathogens on plants. However, the genetic changes wrought in plant populations by the selective effects of pathogens have a reciprocal effect on pathogen populations. This side of the co-evolutionary interaction between plants and their pathogens is considered in Chapter 8. Here, a great deal of emphasis is placed on studies of agricultural plant pathogens as knowledge of those restricted to wild plant communities is very limited.

The final chapter (Chapter 9) highlights gaps in our knowledge and makes suggestions as to how these may be rectified. This is followed by a short glossary of the main plant pathological terms used in this book.

2

The effects of pathogens on individual plants

The range of pathogenic organisms found on plants is enormous. Each of the three major groupings of plant pathogen – fungi, bacteria and viruses (including mycoplasmas and virus-like organisms) – is composed of a large number of species that produce a wide variety of visible and latent symptoms on hosts. A fundamental observation concerning host–pathogen interactions is that not all plants are susceptible to all plant pathogens. However, most are prone to attack by more than one pathogen. These may attack at quite different stages of the plant's life cycle with markedly different consequences for the plant.

There is no simple relation between the loss of a given amount of host tissue to a given pathogen at one stage in the life cycle and the loss of an equivalent area to a different pathogen at some other stage in host development. As a consequence, in order to understand the ways in which pathogens may affect the size and structure of plant populations, we need to appreciate the diverse ways in which pathogen-induced damage is translated into the essential features of the evolutionary performance of host plants – their fecundity and longevity.

Pathogen activity and the life cycles of plants

The basic features of the sequence of developmental events that occur during the life cycle of a plant can be summarized into four main phases (Figure 2.1). The first phase represents dispersed seed that is held in a dormant state in the seed bank. Individual seeds may spend a negligible time in this phase or may remain in it until death. The second phase in the life cycle begins when this dormancy is broken and seeds begin to germinate. Seedlings that successfully withstand the dangers inherent in the early recruitment process become established plants, many of which complete their development, flower and set seed. The final stage in the life cycle covers the events that occur between seed maturation and dispersal.

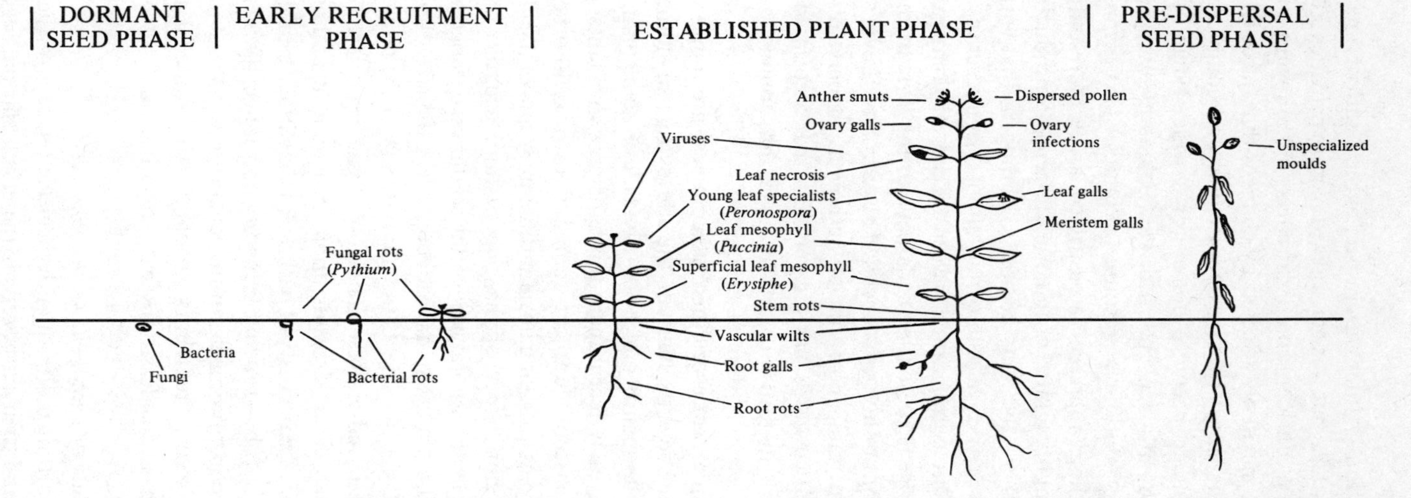

Figure 2.1 A stylized life cycle of a plant showing the four major phases of growth, the different niches provided for pathogens and a representative range of the organisms involved.

Plant species obviously differ with respect to the length of time they remain in each of these stages but this generalized picture provides a useful model on which to base a consideration of the effects of pathogens on their hosts.

Pathogens are inevitably affected by changes that occur during the life of a plant. As the size, structure and architectural complexity of plants increase they tend to become 'home' for a wider range of pathogens. Indeed, from the pathogen's point of view, plants truly represent a constantly changing multi-niche resource (Figure 2.1). Germinating seeds and young seedlings represent a restricted array of resources and are afflicted by a similarly limited group of pathogens. Reproductively mature plants, on the other hand, may be host to pathogens that variously affect parts of the flower or developing fruit, shoots, stems, leaves or roots, of varying ages. This correlation between structural complexity and the number of pathogens present extends beyond the level of the individual species to life forms of hosts in general. Thus Strong & Levin (1979) found that the average number of fungal pathogens associated with North American tree species (15) was greater than that associated with shrub species (7), which in turn was greater than that found on herbaceous species (5).

The relative importance of different ecological groupings of pathogen also varies from stage to stage in the life cycle. Air-borne pathogens responsible for the majority of foliar diseases of adult plants have little effect during the early recruitment phase of growth. Plants at this stage are particularly susceptible to the depredations of soil-borne pathogens. Similarly, pathogens attacking hosts at different stages of the life cycle have substantially different effects on the longevity and fecundity of their hosts. By rapidly overwhelming and killing small seedlings, pathogens that attack plants early in the life cycle may directly affect the size of a developing population. Whether this affects the ultimate size of the adult population or simply removes individuals that would have been lost due to 'self-thinning' (see Chapter 1) depends upon the extent of pathogen-induced losses. In contrast, the majority of pathogens that attack established plants rarely kill them but, by reducing fecundity and longevity, may affect the size and genetic constitution of future generations.

Interactions occurring between plants and their pathogens at each stage in the life cycle are considered in greater detail below.

The effects of pathogens on dormant seed

An important but frequently neglected aspect of the interaction between plants and their pathogens is the effect that pathogens have on individual seeds occurring in the soil. In most, if not all, plant species a very large proportion of the seed produced in one generation fails to give rise to seedlings in the following generation. Without doubt, many of these seeds are lost through the depredations of pathogenic organisms.

Viable seed entering the soil may be in either an active or a dormant physiological state. Physiologically active seed may germinate immediately (see p. 12) or may be forced into a dormant state. Dormant seed fails to respond to apparently suitable environmental conditions and may remain in a viable, but inactive state for considerable periods of time. This difference in the degree of physiological activity occurring in seeds present in the active and dormant seed banks is likely to significantly affect their vulnerability to pathogen attack (Burdon & Shattock, 1980).

Our knowledge of the causes of decay of dormant seeds is negligible. Little useful information can be derived from agriculturally oriented post-harvest studies (reviews of aspects of this literature are to be found in Barton (1961) and Christensen (1978)) as the storage conditions used there are quite unlike those found in the soil. Despite this it seems reasonable to argue that both externally and internally borne seed pathogens will tend to reduce the long-term viability of dormant seed. Surface-contaminating fungi or bacteria may be responsible for the death of seeds directly through necrotic action or indirectly through the production of toxic metabolic wastes. On the other hand, internally borne pathogens (for example, the loose smut fungi *Ustilago* spp. (Mathre, 1978), and many viruses (Shepherd, 1972)) are highly specialized organisms whose effects are likely to be restricted to increases in the general level of metabolic activity in seeds. In turn, this may lead to a more rapid loss in long-term viability.

The effect of soil-borne pathogens on dormant seed is also poorly documented. Potentially, the low physiological activity shown by dormant seed results in individuals whose interactions with the surrounding environment are very limited. This may render dormant seeds virtually 'invisible' to many soil-borne pathogens, thus minimizing losses at this stage (Burdon & Shattock, 1980). However, seeds may remain in a dormant state for many years. As they age there is a gradual decline in the integrity of cell membranes, tight control over the level of metabolic activity is often lost and the general physiological state of the individual deteriorates. An almost inevitable consequence of these changes is a loss of nutrients into the surrounding environment. This will stimulate the growth of fungi and bacteria in the immediate vicinity and hence increase the risk of pathogen attack.

The extent to which physiological deterioration may progress before it affects the viability of seeds is not known. Equally, we do not know whether pathogens that invade dormant seed in response to these changes are attacking seeds that are still capable of germination. Seed stored for prolonged periods under laboratory conditions often shows marked losses of viability (Barton, 1961). That stored under sub-optimal conditions rapidly declines in viability, speed of germination and general seedling vigour (Harrison, 1977). It may be that soil-borne pathogens, preferentially attacking deteriorating seed in the dormant seed bank, are simply accelerating the loss of those individuals rather than destroying seeds that are still in peak physiological condition.

The effects of pathogens in the early recruitment phase

Seeds in the non-dormant seed bank imbibe water and begin to germinate in response to suitable environmental conditions. Imbibing is accompanied immediately by a high rate of exudation of soluble organic and inorganic materials, including a variety of sugars and amino acids (Rovira, 1965; Lynch, 1978). Once rehydration of the seed is complete the rate of exudation falls substantially but for the rest of the life of the plant organic compounds continue to be lost from the roots into the surrounding soil. Exudates from seeds and the roots of young seedlings are a major source of nutrients for many micro-organisms in the soil. Exudates are of particular importance to soil-borne pathogens, as they stimulate the germination of resting spores, induce growth towards the root and generally increase the inoculum potential of the pathogen (Burdon & Shattock, 1980). These effects have been demonstrated on many occasions. For example, Zentmyer (1961) observed the differential accumulation of zoospores of *Phytophthora cinnamomi* (root rot) along young avocado roots and Dukes & Apple (1961) noted a similar phenomenon with respect to zoospores of *P. parasitica* and tobacco roots. The importance of exudates in increasing the susceptibility of plants at this stage of growth has been confirmed in a study involving resistant and susceptible cultivars of soybean. Seedlings of cultivars susceptible to pre-emergent seed and seedling rots caused by *Pythium ultimum* and *P. debaryanum* exuded more than twice the quantity of soluble carbohydrates than did seedlings of resistant cultivars (Keeling, 1974).

The commonest pathogens to attack germinating seeds and young seedlings are probably those responsible for a range of 'damping-off' diseases. Infection by these relatively unspecialized fungal pathogens (for example, *Rhizoctonia solani* and various *Pythium* species) is characterized

Figure 2.2 *Pythium*-induced damping-off of cress seedlings, showing an early stage in the development of disease foci.

by the collapse and death of individual hosts as a result of damage to the root and hypocotyl (Figure 2.2). During the first few days after emergence most individuals are highly susceptible to attack (Chi & Hanson, 1962; Bateson & Lumsden, 1965) but this vulnerability changes rapidly and within 2 or 3 weeks of emergence most seedlings are resistant. Typical results are those of Laviolette & Athow (1971) who found that 100% of 8-day-old soybean seedlings were killed by *Pythium ultimum* but that only 6 days later less than 10% of infected individuals succumbed. Because of these changes in susceptibility, the level of occurrence of damping-off disease can be likened to a 'race' between the rate at which the pathogen mobilizes to the attack and the rate of seedling development (Leach, 1947; Webster et al., 1970). The identity of the most likely winner in any particular host–pathogen encounter is strongly affected by environmental conditions, particularly temperature and humidity. Leach (1947) showed that at any particular temperature the proportion of plants killed prior to emergence could be predicted from the ratio of the linear growth rate of the pathogen to the speed of seedling emergence. A similar ratio probably exists linking the proportion of plants killed after emergence to the growth rate of the pathogen and the rate of tissue maturation.

Pathogens that cause damping-off diseases are amongst the most ubiquitous in the soil and are frequently responsible for the almost complete destruction of seedling stands. However, because diseased hosts often disappear totally within a few days of infection, the effect of these pathogens on the size of seedling populations is generally underestimated.

Seed-borne pathogens also influence the success of seedling establishment. Pathogens carried on the seed coat are in an ideal position to attack the seed as soon as it starts to germinate. Specialized seed-borne pathogens that rely on successful seedling establishment for their own survival frequently reduce the vigour of developing seedlings as do other systematic diseases that gain entry at this stage of growth (for example, *Ustilago avenae* (loose smut of oats)). These reductions in vigour may increase the susceptibility of individuals to attack by other non-specialized pathogens or reduce the ability of infected individuals to compete with others developing around them. Reductions of this nature have been shown in barley seed infected with *Ustilago nuda* (loose smut) (Mathur & Hansing, 1962; Doling, 1964).

The effects of pathogens on established plants

The effects of pathogens on established plants are generally quite different to their effects on germinating seedlings. Not only do diseases of established plants tend to have their main effect through reductions in the growth, vigour and reproductive output of infected individuals, but the greater architectural complexity of the established plant over the emerging seedling results in a greater array of potential niches for pathogens. The root system is susceptible to attack by a variety of soil-borne organisms while the stems, leaves and floral organs all support a further diverse array of parasites (Figure 2.1). Pathogens occupying these niches produce a wide range of morphological effects, ranging from the killing of affected hosts, through various levels of stunting and general reductions in vigour, to apparently insignificant reductions in leaf area or root volume.

Some pathogens reduce or prevent seed production by directly attacking floral organs or developing embryos. *Ustilago* species (the cause of inflorescence smut of a wide range of grasses and members of the Caryophyllaceae), *Pyrenophora seminiperda* (the cause of seed primordia death in *Bromus* species (Neergaard, 1977)) and *Claviceps purpurea* (the cause of ergotism of grasses) all directly reduce host fecundity in this way. Various viruses may induce male sterility in normally self-fertile plants or reduce the vigour of pollen (Mathre, 1978). Other pathogens may affect both the quality and quantity of seed produced by infecting leaves, stems or roots and thus reducing or redirecting the distribution of energy within infected plants. The extent to which such infections affect reproductive performance is strongly influenced by the severity, timing and duration of disease attack. This is considered in some detail later in this chapter but certainly may range from virtually imperceptible effects to almost complete suppression of reproductive activities. Finally, yet other pathogens, such as a range of vascular wilts, may kill adult plants.

Ideally, these effects can all be summarized in a measure of the effect of the pathogen on the fecundity and longevity of the host. However, even at this stage of plant growth, pathogens vary in their degree of visibility to the casual observer. The presence of many pathogens, for example those that produce spreading necrotic lesions on stems and leaves, is apparent within a few days of infection. Other pathogens, however, such as smut fungi, may invade the host at various stages in the life cycle but are not obvious until flowering.

A particularly intriguing group of these 'hidden' pathogens is composed of those that alter host growth patterns in such a way that they actually promote increased longevity, although this may only be apparent in some environments and is often at some cost to fecundity. Perhaps the best known examples of this type of phenomenon are found in diseases like 'choke' of various grasses caused by *Epichloe typhina*. This pathogen causes parasitic castration of its host by inducing abortion of the inflorescence at a very early stage in development. Although this totally prevents seed production, infected plants produce many more tillers than uninfected ones (Bradshaw, 1959).

Similar changes in the growth pattern of grasses have been noted following infection by various viruses. *Lolium perenne* plants infected with barley yellow dwarf virus are shorter and vegetatively more vigorous than healthy individuals. As a consequence, diseased plants are at a selective advantage when defoliation is frequent but at a disadvantage when it is infrequent (Catherall, 1966).

Pathogen infections that produce no obvious symptoms but at the same time alter the longevity and reproductive output of infected hosts pose considerable potential problems for field-based studies. In such studies, differences observed between plants may not reflect genetic variation at all but simply whether or not individuals are infected!

Pathogen-induced pre-dispersal seed damage

In contrast to the many well documented cases of large scale pre-dispersal seed losses caused by animal predation (see review of Janzen, 1971), reports of losses due to pathogen attack are infrequent. This probably reflects the effects of the dry and near-sterile conditions in which many seeds are normally held prior to their release from the parent plant. However, pre-dispersal pathogen damage certainly does occur. In agricultural grain crops warm, moist weather after heading may favour heavy infestations of relatively non-specific fungi like *Alternaria*, *Cladosporium* and *Fusarium* species. Generally, as the grain matures and its moisture content falls, the activity of these pathogens is halted (Christensen, 1978). However, losses may occur when environmental conditions particularly favourable to pathogen development also substantially delay harvesting.

Losses in seed viability, while expressed at the early recruitment phase of the life cycle (Figure 2.1) may be related directly to the environmental conditions under which seed maturation took place in the previous generation. Thus Tenne *et al.* (1974) have shown that the viability of soybean seeds is lowered in areas where temperature, rainfall and humidity are conducive to the infection of pods by a variety of bacterial and fungal pathogens.

Although there are one or two reports of pathogen-induced pre-dispersal seed losses in non-agricultural plants, losses of this nature are generally restricted to agricultural species where retention of seed on the plant aids continued disease development. In non-crop plants this character is relatively rare and seed is dispersed as soon as it is mature. Moreover, even in those species that do retain seeds for prolonged periods these are generally stored in woody cones or capsules that provide protection against the physical environment. In either of these circumstances, the opportunity for substantial pre-dispersal exploitation of seeds by pathogens is strictly limited.

The physiological consequences of disease

Diseased plants are not just smaller, less fecund replicas of their healthy counterparts. Behind the seemingly simple changes in longevity and fecundity that occur as a result of pathogen attack lie a broad array of changes to the general physiology and metabolism of infected host individuals. Pathogen attack that does not kill plants immediately often results in substantial reductions in the rate of net photosynthesis, increases in the rate of dark respiration and marked alterations in the pattern of assimilate partitioning and nitrogen metabolism. Ion uptake and transport and water relations may also be affected (Walters, 1985).

Because of the potentially complicated and far-reaching effects of pathogens on plant growth, it is useful to consider the ways in which these disturbances occur. This highlights and helps to develop a better understanding of a major problem confronting an assessment of the role of pathogens in natural plant communities – that of how to relate varying levels of disease on a variety of different plant parts to changes in the relative fitness of individuals.

While this question is relevant to all pathogens, most studies of the effects of disease on host plant physiology have investigated interactions between foliar biotrophic pathogens and their hosts. This emphasis is not unexpected, as the immediate physiological consequences of disease induced by necrotrophic and biotrophic pathogens are very different. Necrotrophic pathogens gain all their nutritional requirements from recently colonized dead host tissue and, as a consequence, have relatively little direct or immediate effect on the physiology of the rest of the host plant. While infections by such pathogens may ultimately lead to marked shifts in resource allocation patterns, these tend to reflect the host's response to a significant decline in effective photosynthetic area or root volume. In contrast, foliar biotrophic pathogens establish long-term associations with living host cells and, in so doing, may have profound effects on host plant metabolism.

Changes in carbon allocation patterns

Within a short period of time after infection, foliar biotrophic pathogens may induce considerable changes in the pattern of assimilate distribution amongst the various competing energy sinks of healthy plants. Using radio-actively labelled carbon dioxide ($^{14}CO_2$) Livne & Daly (1966) were able to show that the amount of newly fixed carbon being exported from the first leaves of *Phaseolus vulgaris* was substantially reduced if leaves were infected with the pathogen *Uromyces appendiculatus* (bean rust) (Figure 2.3). In healthy plants, 50% of radio-actively labelled assimilates had been transported to other parts of the plant, most notably the stem and root system, within 5 h of exposure of the primary leaves to $^{14}CO_2$. In contrast, in diseased plants nearly all assimilates (98%) were retained in the exposed leaf. Moreover, the presence of this diseased leaf actually acted as a net energy sink, accumulating assimilates fixed elsewhere in the plant. Similar accumulations of assimilates have been observed in infected leaves in a range of other host–pathogen combina-

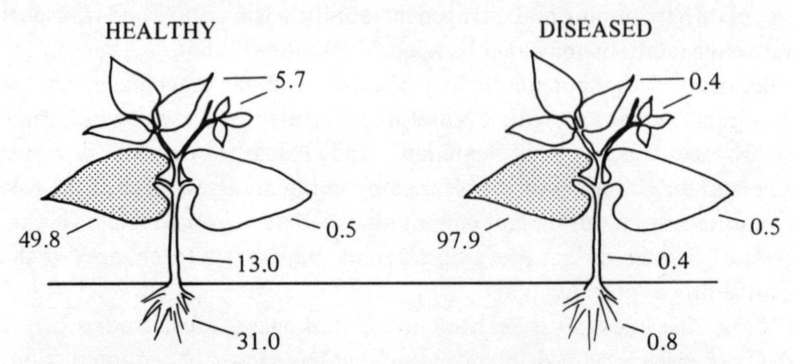

Figure 2.3 Changes in the pattern of carbon translocation from healthy and diseased unifoliate leaves of *Phaseolus vulgaris* infected with *Uromyces appendiculatus*. Changes were determined 5 h after exposure of unifoliate leaves to $^{14}CO_2$ (shaded leaf). The values shown are percentages of the total ^{14}C activity detected. (Redrawn from Livne & Daly, 1966.)

tions, for example, wheat infected with *Puccinia striiformis* (stripe rust) (Siddiqui & Manners, 1971), radish with *Albugo candida* (white rust) (Fric, 1975) and potato with leaf roll virus (Watson & Wilson, 1956; Matthews, 1970); accumulation was also observed in cabbage roots infected with *Plasmodiophora brassicae* (club root) (Mitchell & Rice, 1979). In some cases such disease-induced changes in assimilate distribution patterns have subsequently been expressed as changes in the relative dry weights of roots, shoots, leaves and reproductive structures. In other cases, little change has been observed.

In a study of the effect of leaf roll virus infection on potatoes, Watson & Wilson (1956) found that there was a substantial difference between healthy and diseased individuals in the proportion of resources allocated to tuber production (Figure 2.4). Not only were infected plants much smaller than healthy ones throughout the entire period of growth (50%) but, even by the nineteenth week, less than half of the total weight of diseased plants was found in tubers. A similar fraction was present in the aerial parts of the plant. In healthy individuals, approximately 80% of resources were found in tubers and less than 17% in stem and leaves. In contrast, Ben-Kalio & Clarke (1979) found no noticeable difference in the relative proportion of resources allocated to the roots, stems and leaves of *Senecio vulgaris* heavily infected with *Erysiphe fischeri* (powdery mildew) (Figure 2.5). In this case, measurements ignored the allocation of resources to reproductive structures. However, these were present from the eighth week and, by the tenth week, were three times more numerous on

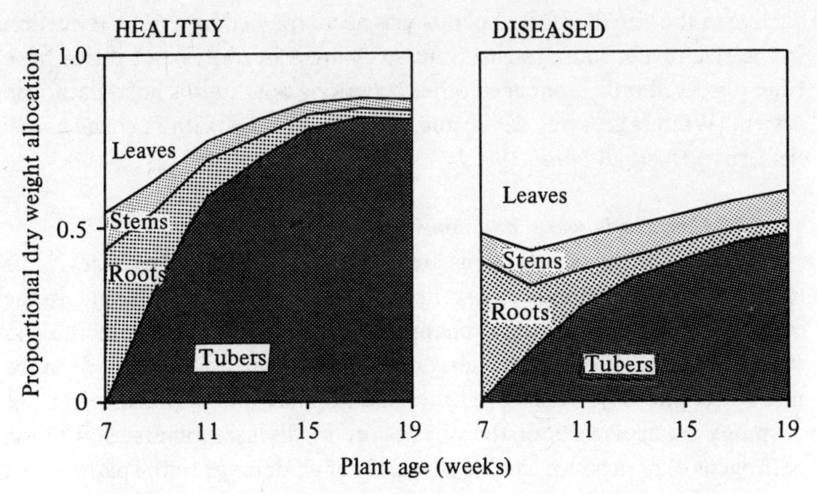

Figure 2.4 Resource allocation patterns (dry weight) in healthy potato plants and those infected with leaf roll virus. Data are for the 12-week period of growth prior to maturity. (Derived from data of Watson & Wilson, 1956.)

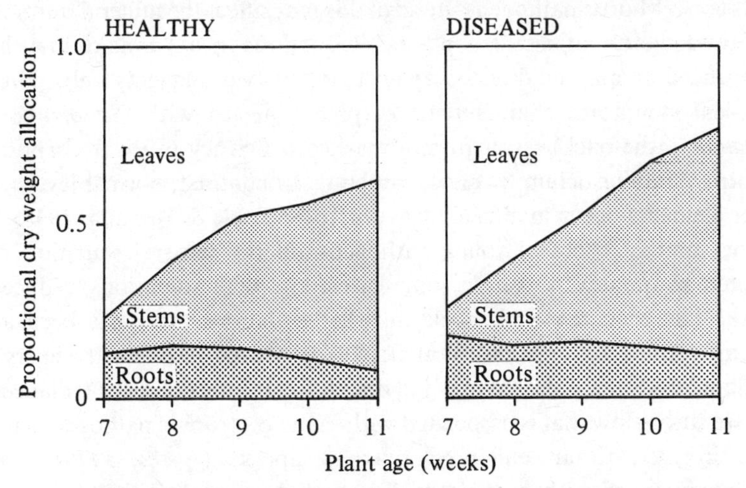

Figure 2.5 Resource allocation patterns (dry weight) in healthy *Senecio vulgaris* plants and those heavily infected with *Erysiphe fischeri* (powdery mildew). (Derived from data of Ben-Kalio & Clarke, 1979.)

healthy than infected plants. The inclusion of these differences would have produced small, but significant, changes in resource allocation patterns. Similarly minor changes were found in a study of the effects of light infections of *E. graminis hordei* (powdery mildew) on the growth of barley (Brooks, 1972). When infections were severe, an overall 16%

decline in the size of infected plants was accompanied by a 66% reduction in the size of the root system! Similar changes in root:shoot ratios have been observed on a number of other occasions both in this host–pathogen system (Walters & Ayres, 1981) and in barley infected with *Puccinia hordei* (leaf rust) (Ahmad *et al*., 1982).

Changes in water and mineral nutrient status

The effects of pathogens are not restricted solely to changes in the total amount of carbon fixed and its distribution within host plants. In the case of foliar pathogens, these changes, further exacerbated by increase in the rate of respiration (particularly dark respiration) and reductions in the rate of net photosynthesis of infected tissues, may induce profound effects on many other aspects of the physiology of diseased plants. Soil-borne pathogens may produce similar effects through damage to the plant's root system.

Reductions in the size or rate of increase of root systems often result from declining assimilate translocation from the shoot or the necrosis of roots by soil-borne pathogens. In turn, this may affect the mineral nutrient and water status of diseased plants. Reductions in ion uptake may be generalized or may be directed against one or two nutrients only. Thus little-leaf symptoms characteristic of pines infected with *Phytophthora cinnamomi* (die-back) result from a marked deficiency in the absorption of nitrogen and calcium by diseased plants. In contrast, normal levels of other elements occur in affected leaves (Roth, Toole & Hepting, 1948).

The indirect effect of foliar pathogens on the mineral nutrition of infected plants is variable. In some cases, and with some ions, reduced uptake has been reported, while in other situations there has been an increase (Ahmad *et al*., 1982; Walters, 1985). Moreover, while the energy available to the root system may fall due to foliar infections, the enhanced transpiration flow that is associated with many biotrophic pathogens may lead, in fact, to an enhanced nutrient uptake (Ayres, 1978). The consequences of such a variety of potential changes may be further complicated by the accumulation of mineral nutrients in diseased tissues.

Most foliar, root and stem diseases alter the water relations of plants in such a way that some or all of the aerial parts of affected hosts are particularly prone to suffer a water deficit. This may occur as a result of accelerated water loss (many biotrophic foliar diseases), disruptions to the transpiration flow (vascular wilt diseases) or destruction of the root system (soil-borne diseases). Paul & Ayres (1984) found that drought combined with the occurrence of the pathogen *Puccinia lagenophorae* (rust) on

Senecio vulgaris had additive deleterious effects on plant growth. During the sporulation of many biotrophic fungal pathogens (for example, *Puccinia* spp.) leaf and stem cuticles are ruptured and the plant is deprived of control over water loss. In other cases, pathogens may inhibit stomatal movement and produce a similar lack of control (Ayres, 1978). In field situations where periods of drought are not uncommon, such an interaction between the occurrence of disease and water loss may greatly reduce the survival and reproductive output of affected hosts.

Detailed expositions of the wide-ranging ramifications of all these effects on the general physiology and metabolism of diseased plants are to be found in several recent reviews (for example, Daly, 1976; Ayres, 1978, 1981; Huber, 1978; Kosuge, 1978; Pegg, 1981; Walters, 1985).

The relation between disease severity and host fecundity

Considering the diverse ways in which plant diseases may affect the physiology of their hosts, it is not surprising that they may also have substantial effects on the fecundity of individual plants. The extent to which this occurs depends upon a range of factors including the severity, timing and duration of disease relative to the growth stage of the host, and on the particular organ infected or, in the case of some foliar infections, on the particular leaf involved. Without doubt, equal areas of diseased tissue do not necessarily imply an equivalence of ultimate effect! For this reason the relative effects of the timing and duration of disease are of particular importance in non-agricultural plant communities where pathogens are often present at low frequency for substantial periods of time.

In the broadest sense, there is a general positive correlation between the severity of disease and its effect on the reproductive capacity of the host. Thus greater disease severity can generally be equated with greater damage. Unfortunately, however, because the effect of pathogen-induced damage is linked to the growth stage of the host, relations between disease severity and host reproduction are rarely simple. Both the timing and duration of disease occurrence are critical variables that greatly affect the overall severity of disease to an individual plant.

The timing of disease occurrence

For a wide range of host–pathogen combinations, equal amounts of disease occurring at different stages in plant growth may have markedly different effects on the final reproductive performance of hosts. Brooks (1972) and Scott & Griffiths (1980) showed that severe, early attacks of *Erysiphe graminis hordei* (powdery mildew) were more damaging to barley than late ones. Early attacks mainly affected fecundity through reductions in the number of fertile tillers although, in some cases, reductions in grain size and numbers of grain per ear were also noted. Most losses from late attacks, on the other hand, were caused by reductions in grain size. A similar pronounced effect of early, severe disease attacks has been found in the interaction between the rust pathogens *Puccinia coronata* (crown rust) and *P. graminis avenae* (stem rust) and the wild oat species *Avena barbata* and *A. fatua*. In all four pair-wise combinations of host and pathogen, early attacks resulted in significant reductions in the number of fertile tillers and average seed size. Attacks that did not begin until heading had little apparent effect (Burdon, 1982 and unpublished data).

A particular problem confronting studies of this type, and in particular those based on observations of plants growing in the field, is that the effects of the timing of disease occurrence are readily confounded by differences in the duration and overall severity of disease. These problems were minimized by Scott & Griffiths (1980) in a carefully monitored experiment involving the judicious use of fungicides to control the occurrence of *Erysiphe graminis hordei* on barley. Their results confirmed the results of other less controlled studies. When epidemics began early and continued to plant maturity, yield losses of more than 50% resulted from a marked decline in the number of fertile tillers per plant and, to a lesser extent, to reductions in the size and number of seeds per ear (Figure 2.6). The effects of early epidemics that started before tillering but which were later controlled were restricted to changes in tiller number. In contrast, most of the reduction in grain yield (24%) resulted from epidemics that began after the end of tillering and was caused by a fall in the size and number of seeds per ear.

Marked reductions in the number of fertile tillers are probably a fairly typical response of most graminicolous hosts to early, severe attacks of disease. They seem to be directly linked to reduced assimilate production during the tillering phase of plant growth. Similarly, reduced grain size is a common feature of late disease epidemics and may well reflect a decline in the post-anthesis photosynthesis that normally contributes significantly to grain filling. The particular importance of the flag leaf and the head itself in this process has been shown in a study of the effects of *Septoria nodorum* (glume blotch) on wheat (Bronnimann, 1968). By inoculating the head, flag, second and third leaves separately and in various combinations, Bronnimann demonstrated the relative contribution that infection of each of these organs made to a 65% reduction in seed size (Figure 2.7). Damage to the head and flag leaf was of near-equal importance, causing 80% of

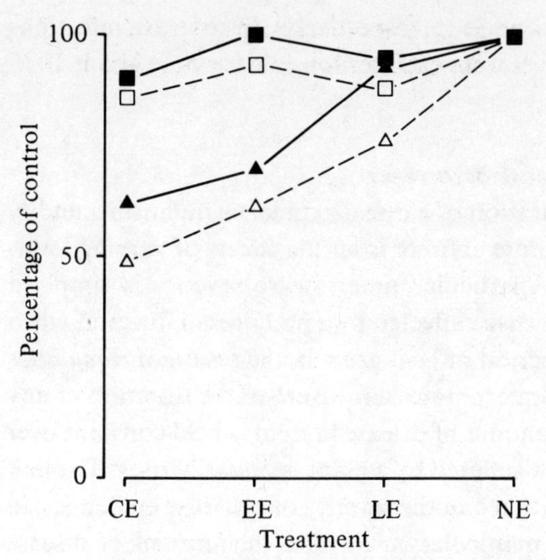

Figure 2.6 The effect of the timing of epidemics of *Erysiphe graminis hordei* (powdery mildew) on various aspects of the reproductive potential of *Hordeum vulgare*. CE, control epidemic: plants infected early and disease allowed to develop until harvest; EE, early epidemic: plants infected at the same time as CE but disease development stopped prior to appearance of first node; LE, late epidemic: plants infected at appearance of first node, disease allowed to develop until harvest; NE, no epidemic. ▲, number of fertile tillers per plant; □, number of grains per tiller; ■, weight of 1000 grains; △, grain yield per plant. (Derived from data of Scott & Griffiths, 1980.)

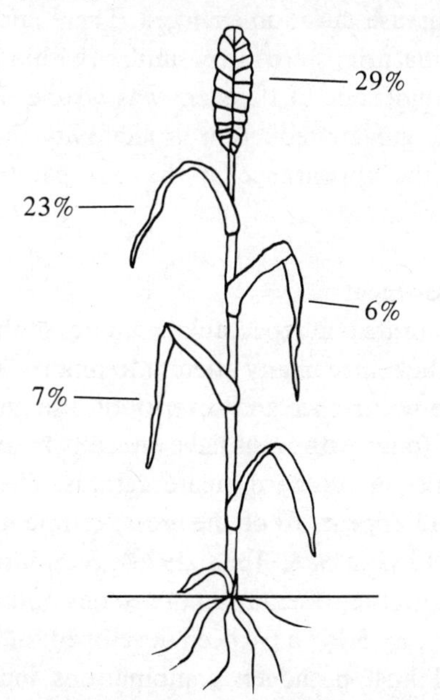

Figure 2.7 The apportionment of a 65% loss in grain size due to the occurrence of *Septoria nodorum* (glume blotch) on wheat. (Redrawn from Neergaard (1977), based on data of Bronnimann, 1968.)

the decline in seed size (45% and 35%, respectively). In contrast, infections on the second and third leaves were each responsible for only about 10% of the reduction.

The duration of disease occurrence

The effect of the duration of a disease epidemic on host fecundity is almost impossible to separate entirely from the effects of varying levels of disease severity. At any particular time disease severity is simply a measure of the area of plant tissue affected by a pathogen. However, when this is extrapolated over a period of host growth, the resultant *cumulative* disease severity necessarily incorporates a measure of the duration of any disease attack. Even if the amount of disease present is held constant over time, the cumulative amount suffered by a plant necessarily rises. The link that exists between the duration and the severity of a disease epidemic can only be broken by careful manipulation of both the amount of disease present and the duration of the infection. Even then the issue is further confused if a minimum level of disease must be attained before damage ensues. In a study of barley infected with *Erysiphe graminis hordei* at six different phases of growth, Scott & Griffiths (1980) found that increasing durations of disease attack resulted in increasing cumulative values of disease severity. This was ultimately reflected in reductions in all aspects of reproductive output. However, in two consecutive treatments more direct evidence for the role of disease duration emerged. There, although cumulative disease severities at maturity were fairly similar (within 20%), plants inoculated when the second node of the stem was visible (Feekes growth stage 7) suffered a 35% greater reduction in fecundity than did those inoculated shortly after the appearance of the last leaf (Feekes growth stage 9).

Problems in disease assessment

The ability to obtain accurate and repeatable estimates of the level of disease on host plants lies at the centre of any attempt to link reductions in fecundity to variations in the occurrence and severity of disease. Over the years, disease assessments of foliar pathogens have typically been made using a range of descriptive scales or standard area diagrams. The latter approach is now widely used and appears to be the most reliable method of disease assessment in the field (James & Teng, 1979). Standard area diagrams or keys provide a pictorial representation of a series of different levels of disease severity (Figure 2.8). Keys have been developed for a wide variety of agriculturally based host–pathogen combinations including

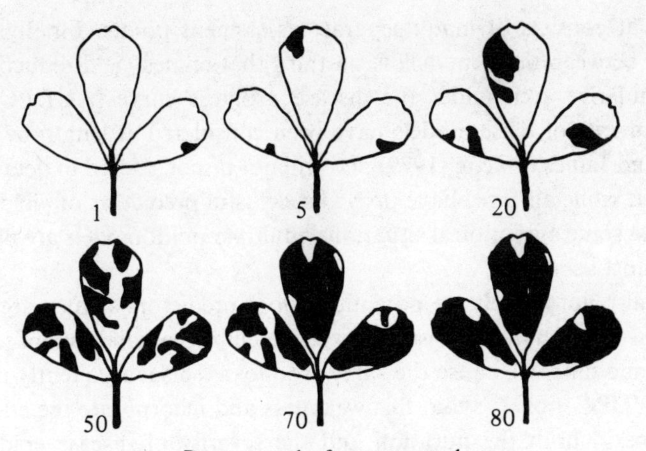

Figure 2.8 Disease assessment key for yellow leaf blotch of lucerne caused by *Leptotrochila medicaginis*. The key is used by visually comparing diseased leaves with these diagrams and hence estimating the percentage of leaf area affected. (Redrawn from James, 1971.)

Erysiphe spp. (mildews), *Puccinia* spp. (rusts) and *Rhynchosporium secalis* (scald) on cereals, *Phytophthora infestans* (late blight) on potato and *Bremia lactucae* (downy mildew) on lettuce (James, 1971; Agricultural Development and Advisory Service, 1976). The amount of disease occurring in a crop is visually estimated by comparing disease levels on a random sample of plants with these standardized diagrams. For systemic diseases (for example, those caused by wilt fungi and most plant viruses) or soil-borne diseases which may totally destroy young seedlings, a more relevant measure is obtained by regarding each infected host individual as a single unit of disease.

Disease assessment keys are equally applicable to non-agricultural host–pathogen interactions. In many cases they can be used directly to estimate disease severity in such systems (for example, *Erysiphe* or *Puccinia* spp. occurring on a range of grasses) while in other cases new keys may have to be devised. Similarly, disease assessment keys can be used in studies monitoring disease levels on individual hosts through time, just as validly as to assess disease levels on randomly selected plants.

Once obtained, disease assessment data may be used in a number of mathematical models to predict levels of yield loss in the host plant. Three categories of model have been developed. Critical-point models provide an estimate of the yield loss incurred for any given level of disease at a given time. Multiple-point models estimate losses from a disease progress curve that consists of many separate assessment points. Finally, intermediate between these models lies a third that relates yield reductions to the cumulative area under the disease progress curve (AUDPC). The relative merits of these models have been considered at length by James (1974) and James & Teng (1979) and will not be considered in detail here. However, while all three have proved successful predictors of yield losses in at least some agricultural situations, multiple-point models are likely to be the most accurate.

Critical-point models are potentially poor predictors as they are based on the assumption that all disease curves reaching the same level of severity at the same time will cause the same crop loss. This is frequently not the case. AUDPC models avoid this weakness and incorporate the effects of variations in both the duration and the severity of disease epidemics. However, they do not allow for differences in the relative importance of disease at different growth stages of the host. Multiple-point models overcome all these problems by providing a measure of both the cumulative level of disease and of its severity at particular stages during the growth cycle. They consequently provide a more accurate picture of the relation between disease severity and plant reproductive performance.

Despite the obvious attractions of multiple-point models, from the point of view of studies of host–pathogen interactions in non-agricultural plant communities, these models are not very practical. Multiple-point models rely on accumulated knowledge concerning the physiological response of the host to disease at all stages of growth. This knowledge is important in determining the points at which disease severity is assessed and the relative weighting which each value is given in the final model. In situations where relatively little is likely to be known about the physiological response of the host to disease (most non-agricultural host–pathogen combinations), these models are inappropriate. In contrast, AUDPC models have no *a priori* knowledge requirements. Despite their shortcomings they are, therefore, probably the most practical models for relating the severity of disease to reductions in host fecundity in non-agricultural systems. Moreover, on many occasions AUDPC models have shown very close correlations between the cumulative level of disease and the resultant disease loss (van der Plank, 1963; James & Teng, 1979; Scott & Griffiths, 1980).

In the preceding discussion, most attention has been centred upon the problem of relating observed levels of disease to reductions in the fecundity of annual plants, particularly cereals. However, problems concerning the ultimate relation between disease severity and its effects on reproductive fitness are nowhere more acute than in considerations concerning perennial host species.

In perennial species, pathogen-related damage may affect both fecundity and longevity. Most perennials pass through a prolonged juvenile stage of development before reaching reproductive maturity. During this juvenile phase all available resources are allocated to growth, while in the adult phase these are divided between continued growth and reproduction. Despite this change in resource allocation patterns, damage caused by pathogens at either stage may have both immediate and long-term consequences. By reducing the current and short-term future rate of development of individual plants, such damage may affect: (1) the ultimate size and annual fecundity of individuals; and (2) long-term survival. Damage during the juvenile stage may alter competitive relationships between adjacent plants and lead to the long-term suppression of individuals. Eventually, this will be expressed in reductions in adult size and fecundity. Alternatively, the length of the pre-reproductive stage may increase. Pathogen-induced damage to adult plants may be expressed in an immediate reduction in the current year's reproductive effort or, as occurs in many tree species, in a reduction in that of subsequent years.

Diseases may also affect the survival of perennial species. Reductions in longevity may be cumulative so that the total life span of affected individuals is subtly reduced in comparison to unaffected individuals. However, more rapid death may occur. Once again though, there may be a time delay between the occurrence of disease and subsequent death. Plants weakened by disease are often more sensitive to periods of unfavourable environmental conditions. This has been shown recently for perennial species of *Phlox*. Plants heavily infected with *Erysiphe cichoracearum* (powdery mildew) during the summer and autumn of one year, showed lower survival in following winters than did unaffected individuals (Jarosz & Levy, personal communication).

3

The development of disease in plant populations

In most host–pathogen relationships severe epidemics of disease are the exception rather than the rule. Fluctuations in severity may occur from season to season as a result of changing environmental conditions but, overall, disease remains inconspicuous to the casual observer. The processes involved in the maintenance and development of a pathogen population remain the same, regardless of whether that population is passing through a quiescent phase or one of explosive growth. However, it is during epidemic development that the relative importance of the various different processes is most apparent.

The development of an epidemic

Plant disease epidemics are dynamic processes in which pathogen populations increase greatly in size and spread through populations of susceptible host plants. However, unlike populations of plants and animals, the size of a pathogen population is not readily assessed by direct counting. Individuals are generally too numerous to count or cannot easily be distinguished. As a result, the amount of disease present is commonly used as a measure of the size of pathogen populations. The ways in which this may be measured were considered briefly in Chapter 2.

Epidemic growth curves are typically sigmoidal in shape. Each stage in the development of the epidemic melds into the next to form a single smoothly continuous process (Figure 3.1). Despite this, epidemiologists have found it convenient to recognize three distinct phases (Zadoks & Schein, 1979): (1) an initial *exponential* phase during which pathogen numbers undergo their greatest relative multiplication as the amount of disease (x) increases from a hardly detectable trace to a frequency of approximately 0.05; (2) a *logistic* phase which runs until approximately half of all susceptible tissue is diseased ($x = 0.50$); and (3) a *terminal* phase which sees the epidemic run to completion. This division, however, pays

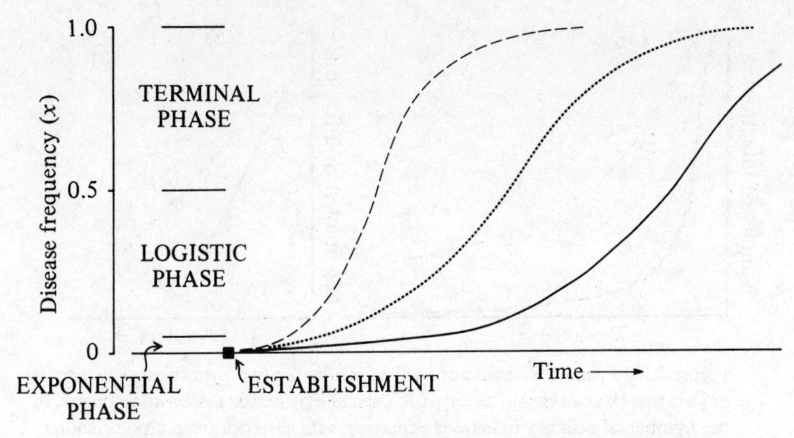

Figure 3.1 The four main phases in the development of plant disease epidemics. The three hypothetical epidemic growth curves illustrate how the length of each phase depends upon the overall rate of epidemic development.

relatively little attention to the initial establishment of a pathogen in a host population. This is of particular interest in a consideration of the maintenance and development of disease in natural plant communities and is included here in an additional phase concerning the establishment of disease.

The establishment of disease in a population

Sources of inoculum

The occurrence of disease in plant communities is patchy. At any given time not all individuals in a population are likely to be infected equally. Even in populations in which all individuals respond identically to a pathogen or which are in the grip of an epidemic, large differences in the severity of disease are often found between adjacent plants.

A wide range of pathogens, including many viruses and biotrophic fungi, rely on the continual presence of living susceptible host material for their growth and survival. These pathogens lack the thick-walled dormant spores that assist others to survive unfavourable conditions in a protected state. In many seasonal environments pathogens lacking these resting stages are likely to become locally extinct and will need to recolonize host stands from nearby or distant refugia once favourable conditions return. On the other hand, necrotrophic foliar pathogens that are capable of survival on dead host tissue and most soil-borne pathogens are likely to survive *in situ* until favourable conditions return and susceptible hosts reappear. This spatial disposition of initial inoculum sources relative to

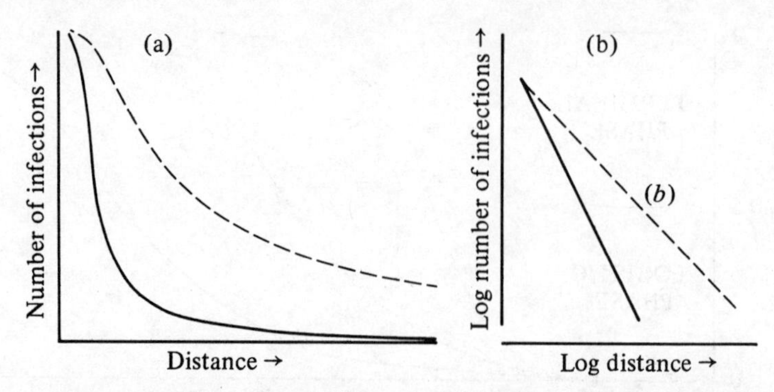

Figure 3.2 Primary disease dispersal gradients for two hypothetical pathogens differing in their dispersal ability. (a) Typical leptokurtic distribution curves of the number of primary infections occurring with distance from a point source of inoculum. (b) The same results plotted on a double log basis showing the resultant linear relations. The slopes (*b*) of these lines are measures of the steepness of the disease dispersal gradients.

new uninfected host populations is of considerable importance in determining the likely pattern of distribution of incoming inoculum and the subsequent distribution of primary infections.

Primary disease dispersal gradients

Disease dispersal gradients are the visually apparent record of inoculum dispersal patterns. They are usually characterized by inverse power relations between the frequency of primary infections and distance from the inoculum source (Figure 3.2a). Close to the inoculum source there is a rapid decrease in the frequency of infections, disease dispersal gradients are obvious and primary infections are distributed in a non-random fashion. With increasing distance the rate at which this decline in infection occurs reduces until little further change in frequency occurs even over relatively large distances. At these distances, gradients are no longer obvious and inoculum and hence primary infections appear to be distributed at random.

Because of their shape, curves of this nature are difficult to compare with one another and are typically converted to a linear form by regressing the logarithm of the number of primary infections against the logarithm of the distance from the inoculum source (Figure 3.2b). The slope (*b*) of the resultant line is taken to be a quantitative measure of the steepness of the disease dispersal gradient (Gregory, 1968).

The shape of disease dispersal gradients is influenced by many factors. Two of particular importance are the size and nature of the inoculum source and the means by which pathogen propagules are normally dispersed (Gregory, 1968). Thus point sources of inoculum (for example, single infected plants) give rise to steeper dispersal gradients than do line or area sources. Similarly, while there are no infallible rules concerning the primary disease dispersal gradients of individual pathogens, in general, wind-borne, splash-borne, soil-borne and contact dispersal provide a sequence of transmission modes characterized by increasingly steep dispersal gradients. On the other hand, pathogens that are transmitted by vectors (for example, viruses or mycoplasmas transmitted by insects or arthropods) show a wide range of infection gradients. Viruses carried by active vectors which remain infective for long periods of time tend to have shallow gradients, while those transmitted by short-lived relatively immobile vectors tend to have much steeper ones (Thresh, 1976). Fungal diseases like Dutch elm disease (*Ceratocystis ulmi*), carried by bark beetles with very limited dispersal ability, also have steep dispersal gradients ($b > 2$), while those transmitted by wide-ranging vectors (for example, *Endothia parasitica* carried by woodpeckers) have very shallow dispersal gradients ($b < 2$) (Harper, 1977).

At an early stage in the establishment of disease in plant populations, differences in the shape of disease dispersal gradients are important in determining both the likely sources of inoculum for uninfected host stands and the distribution of primary infections occurring within these populations. Equally, as we shall see later, as disease develops in the population, dispersal patterns are of considerable importance in determining the pattern of development of disease in the population as a whole.

Initial infection of the host stand

Two distinct patterns of distribution of primary infections are commonly observed in plant populations – random and aggregated (or clumped) (Figure 3.3). Random distributions generally result from inoculum entering a host stand from a distant source (exogenous inoculum; Figure 3.3a). However, while this is almost invariably the case for air-borne pathogens responsible for foliar diseases, random distributions of seed-borne diseases, like some smuts, may result from efficient dispersal of infected seed. This is particularly so in agricultural situations where seed is thoroughly mixed during sowing. Vector-borne pathogens may show clumped patterns of primary infection even though inoculum has originated some distance away from the host stand. In such cases the distribution of incoming inoculum is likely to be strongly influenced by factors which affect the behaviour of vectors. For example, aphid vectors of a variety of plant

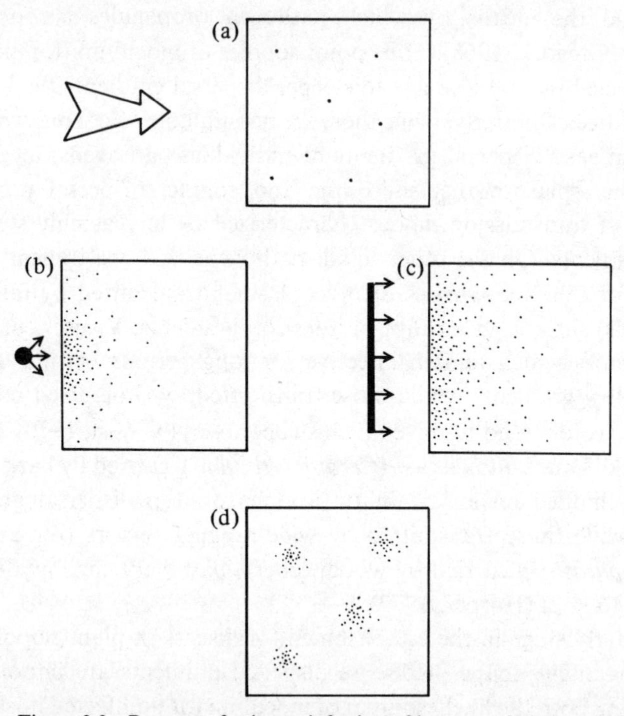

Figure 3.3 Patterns of primary infection of host populations resulting from inoculum originating at various distances from the plant stand. No secondary spread of the pathogen has taken place. (a) Random distribution of primary infections resulting from air-borne inoculum derived from a distant source. (b), (c) Aggregated patterns of distribution of primary infections resulting from inoculum entering the host stand from a nearby point or line source of inoculum, respectively. (d) Aggregated patterns of distribution of primary infections resulting from inoculum originating from within the host stand.

viruses are attracted to land by yellow wavelengths of light such as those reflected skyward from bare ground but are repelled by blue wavelengths which tend to be reflected from a continuous cover of vegetation (A'Brook, 1973). As a result, even when vectors carry pathogen inoculum from a distant source they may land preferentially on isolated plants, producing a non-random distribution of primary infections.

Aggregated or clumped patterns of inoculum distribution are more characteristically associated with inoculum sources near to, or even within, individual host stands (Figure 3.3b,c,d). The actual nature of the distribution pattern depends upon the size and position of the source. Point sources of inoculum occurring in close proximity to a host stand tend to be responsible for fan-shaped patterns of primary infections, the frequency of which increase as the source is approached. Line sources, on the other hand, tend to be characterized by a general increase in the magnitude of incident inoculum on one side of the population as opposed to the other. Such patterns often occur in agricultural situations where viral pathogens spread from common hedgerow plants to young plants in a developing crop.

In contrast to the varied effects of exogenous inoculum sources, endogenous ones almost invariably result in non-random patterns of inoculum deposition and hence the occurrence of primary infections. These sources may be infectious plant debris remaining from past seasons or previously quiescent infections that have recently become active. However, because endogenous inoculum sources actually occur within the host stand, the distribution of primary infections around these sources will be clumped although the spatial disposition of the aggregates themselves may be random or clumped (Figure 3.3d).

The preceding considerations are equally applicable to pathogens invading agricultural crops and wild plant populations. However, the frequency and probability of success of such invasions is likely to differ between the two situations. In most agricultural crops the relatively large size of individual populations, their genetic uniformity and their annual or pseudo-annual habit, generally ensure that once a pathogen establishes primary infections it is likely to remain present for the rest of the life of the crop. This is also true for many wild plant species. For others, particularly those distributed throughout a habitat in small, discrete patches, pathogen extinction may follow local colonization far more frequently. In these host–pathogen interactions the numerical and spatial distribution of host individuals within patches and of patches within an environment is likely to result in, at the most, only a few successful infections occurring in a given patch at a given time. Initial colonization of a host patch by a pathogen may easily fail if premature leaf senescence or a short period of unfavourable environmental conditions prevents effective reproduction of individual pathogen colonies before they die. An analogy of colonization and extinction can be drawn between this situation and that occurring in island habitats (MacArthur & Wilson, 1967). As Janzen (1968) has pointed out with respect to herbivorous insects, individual host plants or even single leaves can be regarded as islands in both a spatial and an evolutionary sense.

The exponential and logistic phases of epidemic development

Disease focus development

The successful establishment of infection in a new host stand represents a single, albeit major, step in the process of disease development in plant populations. Once present, a pathogen must reproduce and spread if it is to survive and its population size is to increase. The spread of a pathogen from primary infections within a host stand to other host plant individuals (focus development) is governed by the same physical processes as those already considered concerning initial establishment. However, from an epidemiological point of view the effects of dispersal for focus development are quite different (Zadoks & Schein, 1979). In particular, it is appropriate to consider the consequences of the effects of different patterns of inoculum dispersal on the way in which disease spreads and increases within a homogeneous host population.

As primary infections become reproductively active, propagules are produced; these disperse, produce new secondary infections and so give rise to the development of a disease focus. As the term implies, a disease focus is simply the site of a local concentration of diseased plants or lesions generally grouped about a primary infection (Federation of British Plant Pathologists, 1973). Given suitable environmental conditions, with each passing pathogen generation the size of the focus will grow and disease will spread further through the host population. The rate at which this occurs is greatly influenced by environmental conditions and the dispersal capacities of the pathogen. It is, perhaps, best illustrated by considering the progress of two hypothetical pathogens that are representative of the two extremes of a dispersal continuum (cases I and III, Figure 3.4).

At one end of the scale (case I), inoculum dispersal is very efficient and primary disease gradients are long and relatively shallow ($b \leqslant 2$). Here, although dispersal efficiency is high most propagules still fall relatively close to the parent lesion. However, individual foci expand rapidly and many secondary foci are established by propagules dispersed well beyond the edge of the expanding primary focus. Over only a few pathogen generations disease gradients flatten as disease becomes common throughout the whole population (Figure 3.4a). The spread of the pathogen rapidly moves from being confined to localized spots of infection to a general eruption across the whole of the host population.

At the other end of the dispersal scale (case III, Figure 3.4), inoculum dispersal is very poor, and primary disease gradients are short and very steep ($b \gg 2$). A cross-sectional view of individual foci after a few pathogen generations shows plants close to the centre to be heavily infected while

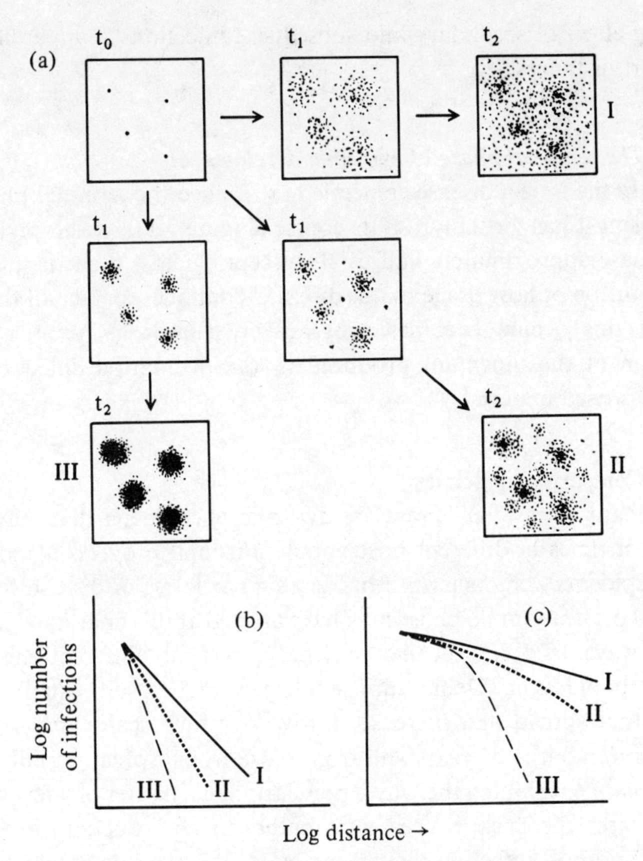

Figure 3.4 The development of disease foci. (a) A diagrammatic representation of the development of disease foci. The three cases represent a sequence, at times t_0, t_1 and t_2, of decreasing efficiency of spore dispersal and provide a visual indication of the extent and rapidity with which individual foci develop and give rise to secondary infections. (b) Primary disease dispersal gradients for the three illustrative cases. The most efficiently dispersed pathogen (I) has the shallowest dispersal gradient. (c) The flattening effect of secondary and subsequent infections on disease dispersal gradients. This effect is most pronounced for the most efficiently dispersed pathogen (I) and least pronounced for the pathogen with the poorest dispersal ability (III).

those only a short distance away are almost, if not completely, free of disease. Disease in the stand as a whole increases slowly as the pathogen spreads as an advancing wave or front through the population. Over time, graphical plots of the logarithm of the number of infections against the logarithm of distance show a change in the shape of dispersal gradients (Figure 3.4). However, compared to efficiently dispersed pathogens the flattening effect of secondary and subsequent infections is apparent only over short distances.

The terminal phase of epidemic development

By the time a disease epidemic has reached the terminal phase of development it has virtually run its course (Figure 3.1). At the beginning of this phase approximately half of all susceptible host tissue is infected. The proportion of host tissue that is diseased continues to rise but the rate at which this occurs becomes progressively slower as an increasing proportion of the inoculum produced is lost in multiple infections of already diseased material.

Comparing epidemics

Plant disease epidemics are dynamic phenomena that may start at different times in different host populations and progress at different speeds. Epidemics characterized by very similar levels of disease severity at the end of the terminal phase may have arrived at this final state in quite different ways. In doing so, they may have markedly different effects on the host population. Debilitating epidemics may result equally from diseases that spread and increase slowly over time scales that may be measured in months or years and from those which spread rapidly from plant to plant enveloping the whole population in a matter of a few weeks. In both cases the basic processes of pathogen reproduction, inoculum dispersal and host infection remain the same. It is the rate at which these processes occur that produces obvious differences in disease severity.

Epidemics can broadly be divided into two categories. Polycyclic epidemics are those in which many generations of the pathogen occur as it spreads from plant to plant in a single growing season. Outbreaks of stem rust (*Puccinia graminis tritici*) of wheat, late blight (*Phytophthora infestans*) of potatoes and blast (*Pyricularia oryzae*) of rice are all examples of polycyclic disease epidemics. Because of their short generation times (10–12 days), massive spore production and reasonably efficient spore dispersal, these pathogens can all increase very rapidly. They may rise in occurrence from below the perceivable threshold to completely destroy host stands within a single growing season. In monocyclic epidemics, on the other hand, once the pathogen has gained entry to a host further spread to other hosts does not occur until the following season. Epidemics of many smuts (*Ustilago* spp.) of grasses and vascular wilts (*Fusarium* spp.) of cotton, banana and a range of other plants are all examples of monocyclic diseases. Epidemics of these pathogens build up slowly as multiple within-season plant-to-plant transmission does not occur.

Even within these broad categories disease epidemics may progress at remarkably different rates due to the interaction of environmental, host and pathogen influences. Consequently, simple assessments of the level of disease severity at a single arbitrary time provide little information concerning the dynamic processes involved or of the likely effect that a pathogen might have on the survival and reproductive capacity of the host population.

Epidemics of plant disease can be measured in two different ways. They may be assessed either directly by measuring the amount of disease present and the rate at which this increases or, alternatively, by measuring the consequences of disease in terms of its effect on the host population. From the point of view of the plant, assessments of the effects of disease are *relatively* simply determined through estimates of the basic parameters of survival and fecundity. The latter parameter, however, may be complicated by trade-offs between the size and total number of individual propagules produced. Changes in these sorts of characters have a direct bearing on the effects of pathogens on the population dynamics of individual host plant populations and are considered in detail in Chapter 6.

Here we are concerned with the problem of how to compare different epidemics directly. As noted previously, because different epidemics start and finish at different times and progress in different ways, comparisons at single points in time may lead to erroneous conclusions. For greater accuracy a means of comparing the rate of epidemic development over a series of consecutive measurements made during the exponential and logistic phases of growth is needed. Such an approach has been pioneered by van der Plank (1960, 1963, 1968) who applied simple mathematical principles to the analysis of epidemiological data. In particular, van der Plank argued that early in the development of polycyclic disease epidemics, nearly all susceptible host tissue is still healthy and the spread and increase of the pathogen is practically unhindered by a lack of susceptible tissue. At this stage the shape of the epidemic curve is exponential* (Figure 3.1) and can be modelled by the equation

$$\mathrm{d}x/\mathrm{d}t = rx \tag{3.1}$$

where x is the proportion of host tissue that is diseased and r is the rate of disease increase.

However, as the amount of disease increases in the host population the

* van der Plank used the term 'logarithmic'; here I follow the mathematically more accurate terminology used by Zadoks & Schein (1979).

possibility of interference and wastage of inoculum due to multiple infection rises and the amount of diseased tissue no longer increases at an exponential rate. Van der Plank made allowance for this by incorporating in equation 3.1 a measure of the proportion of the total amount of susceptible tissue that was still available for infection. Thus,

$$dx/dt = rx(1-x) \tag{3.2}$$

The rate of disease increase or apparent infection rate, r, can then be determined by integration and rearrangement of equation 3.2 so that, in a generalized form,

$$r = \frac{1}{t}\{\ln[x/(1-x)] + c\} \tag{3.3}$$

where c is a constant. Using this basic approach the estimation of r values can be further refined to allow for the continued growth of the host stand during the progress of the epidemic.

With monocyclic disease epidemics, van der Plank showed that this model was inappropriate. During the early phases of such epidemics the rate of disease progress is dependent solely on the number of initial infections per season. Monocyclic epidemics can be modelled by the equation

$$dx/dt = r_s(1-x) \tag{3.4}$$

so that the infection rate, r_s, is given by:

$$r_s = \frac{1}{t}\{\ln[1/(1-x)] + c\} \tag{3.5}$$

The major advantage of van der Plank's approach is that it summarizes the development of an epidemic in a single value, r, which is simply related to the directly measured characters of the amount of diseased tissue present (x) and time. However, the use of the terms $\ln[x/(1-x)]$ or $\ln[1/(1-x)]$ instead of x as a measure of disease is only an effective way of handling data collected during the exponential and logarithmic phases of epidemic development (Figure 3.5). As the epidemic enters the terminal phase van der Plank's models become increasingly inaccurate. Undoubtedly, more complex mathematical models can be devised which predict the behaviour of individual epidemics more accurately (Kranz, 1974). However, it is the basic simplicity of van der Plank's models and their general applicability to an enormous range of experimental results obtained from many different host–pathogen interactions, occurring in a wide range of environments, that make them so useful. A range of apparent infection rates calculated for a variety of disease epidemics is presented in Table 3.1. It

Table 3.1 *Some examples of rates of disease increase found in field situations*

Pathogen	Host	Location		Rate of increase (per unit of host tissue per day)	Reference
Phytophthora infestans	potato	Kentisbere	1942	0.506	Large, 1945
	potato	Darlington	1942	0.461	Large, 1945
	potato	Darlington	1943	0.113	Large, 1945
	potato	Darlington	1944	0.337	Large, 1945
	potato	Dunsford	1944	0.295	Large, 1945
Erysiphe graminis	barley	—	—	0.390–0.452	Burdon & Whitbread, 1979
Tobacco mosaic virus	tobacco	—	—	~0.100	van der Plank, 1963
Fusarium oxysporum	banana	—	—	~0.001	van der Plank, 1963

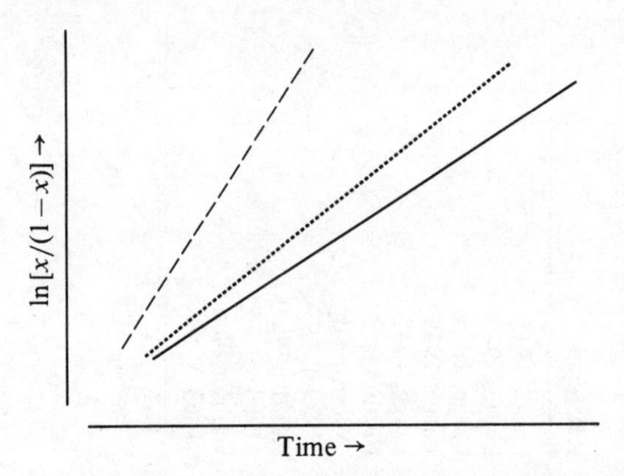

Figure 3.5 The effect of plotting $\ln[x/(1-x)]$ instead of x against time for the three epidemic curves shown in Figure 3.1. The slope of the lines is a measure of the apparent infection rate or rate of increase of the disease.

is important to note that not only do differences occur between different pathogens, but that even for the same host–pathogen combination quite marked differences in infection rates may occur (see *Phytophthora infestans*, Table 3.1).

The effect of plant populations on disease development

Plant disease epidemics do not occur in a vacuum. Rather, they result from a favourable interaction of host, pathogen and physical environmental influences. To this stage we have implicitly assumed that all these conditions have been ideal for the growth of the pathogen and hence the development of an epidemic. In reality, however, this is rarely the case. Physical environmental restraints on pathogen development (for example, host plant nutritional status, temperature, water availability and so forth) are treated in detail in Chapter 5. Here we are interested in the major features of host plant populations which may affect the size of pathogen populations. In particular, what effects do differences in the density of host plant populations and host genotype have on the rate of increase and ultimate size of pathogen populations? These factors may be relatively unimportant in agricultural situations where individual populations tend to be large and of uniform density and genetic constitution. They are of considerable relevance in non-agricultural situations, however, where individual plant populations are often genetically heterogeneous and tend to be fragmented into small colonies or patches of variable density.

Host density and disease severity

The relation between plant density and disease occurrence

The literature concerning the relation between the occurrence of plant disease and host density is confusing as it covers results obtained from single (the majority of cases) and multiple observations of disease in different host stands. In 'short term' situations where neither host nor pathogen has time to reproduce (Burdon & Chilvers, 1982) and observations are invariably taken only once, a doubling of density in one host stand relative to another may produce a variety of different outcomes in regard to the absolute quantity and the relative proportion of plants that are diseased.

If the pathogen is seed-borne the most likely result is for the absolute quantity of disease to double while the proportion of diseased individuals remains the same. If disease results from a limited amount of inoculum introduced by winged vectors, there may be no change in the absolute quantity of disease. However, the proportion of plants infected in the higher-density stand would be half that in the lower-density stand. Finally, in the case of limiting amounts of soil-borne inoculum, the absolute quantity of disease may fall short of doubling and hence the proportion of plants affected would be somewhat reduced (Burdon & Chilvers, 1982). Without doubt these sorts of complications are responsible for the high number of negative correlations that have been observed in short term interactions between host density and the occurrence of viral diseases (Table 3.2). These differences serve to emphasize the importance of multiple determinations of disease severity over time.

Table 3.2 is a summary of studies of the effect of host density on disease occurrence. It is similar to a table previously presented by Burdon & Chilvers (1982) but has been enlarged by the addition of a further 21 records. In all, data from a total of 90 studies relating to 61 different combinations of host and pathogen are represented. Of these studies, 56 provided evidence of a positive correlation and 26 of a negative correlation between host density and disease intensity. Half of the latter relate to diseases caused by viruses while over 85% of the positive correlations were provided by fungal diseases. In the majority of these studies, data were collected only once or twice from only two or three host stands of different densities. Where investigations of a wider range of host densities have been coupled with multiple assessments of disease intensity throughout the development of epidemics, clearer pictures of the relation between these variables have emerged. Over very wide ranges of density (450–12600 plants per m^2), curvi-linear relations between the rate of disease increase

Table 3.2 *Summary of literature references to host density effects on disease severity expressed as the proportion of diseased plants (after Burdon & Chilvers, 1982*[a]*; reproduced, with permission, from the* Annual Review of Phytopathology, *Vol.* **20**. *© 1982 by Annual Reviews Inc.*)

Time scale[b]	Type of pathogen	Correlation between host density and disease severity[c]		
		Positive	Nil	Negative
Short term	Fungus	1 (1)	0	2 (2)
	Virus	1 (1)	0	13 (5)
Medium term	Fungus	47 (36)	7 (6)	7 (4)
	Virus	6 (4)	0	0
Long term	Fungus	1 (1)	1 (1)	4 (2)
	Virus	0	0	0
Totals		56 (41)	8 (7)	26 (13)

[a]In addition to the references given in Burdon & Chilvers (1982) the following citations have been added: Smith & Blair (1950); Mannson (1955); Kammeraad & Brewer (1963); Jenkyn (1970); Gheorghies (1972) ; Steadman, Coyne & Cook (1973); Timmer & Fucik (1975); Huang & Hoes (1980); Dow, Porter & Powell (1981); Inouye (1981); Pataky & Lim (1981); Shukla & Anjaneyulu (1981); Cobb *et al.* (1982); Jennersten, Nilsson & Wastljung (1983); Alexander (1984); Augspurger & Kelly (1984). [b]Short term = no reproduction; medium term = pathogen but no host reproduction; long term = both host and pathogen reproduction. [c]Figures in parentheses give the number of different host–pathogen combinations represented.

or spread and host density have been demonstrated for damping-off diseases caused by pathogens like *Pythium* (Gibson, 1956; Burdon & Chilvers, 1975). For diseases like damping-off, that spread mainly through mycelial growth from plant to plant, such results are not unexpected. Here the curvi-linear relation reflects an inverse linear correlation between the rate of disease increase and the average distance between hosts! For air-borne diseases, the range of densities examined has been much more restricted and, there, linear relations between plant density and disease rates have been identified (Burdon & Chilvers, 1976*a*).

Unfortunately, a large majority of the studies listed in Table 3.2 deal only with situations in which either both host and pathogen do not reproduce (short term time scale) or in which only the pathogen reproduces (medium term). Very few data are available concerning the ecologically most interesting long term time scale. There, because both host and pathogen have had sufficient time to reproduce, observable differences in disease rates may be taken to reflect the reciprocal effects of host on pathogen and pathogen on host. Major complications can arise however. Thus in the interaction between the pathogen *Cronartium ribicola* (blister rust) and its pine hosts (this provides three of the negative long-term fungal values in Table 3.2), the possibility of a simple relation between the density of pines and the occurrence of the pathogen is confounded by the presence of the alternate host (*Ribes* spp.) of this heteroecious pathogen. As the density of pines declines the frequency of *Ribes* spp. in the forest understorey actually rises. This, in turn, results in an increase in the occurrence of disease!

Certainly the long-term interaction of host density and disease intensity is an area of research in which further studies would be extremely valuable. However, as seen in the above interaction, the interpretation of results may be fraught with difficulty due to the possibility of significant changes in host, pathogen and/or environment.

Despite the general positive correlation between disease severity and host density shown in Table 3.2, individual spatially isolated plants have often been observed to be heavily diseased. For example, in Europe, Gaumann (1950) noted a high level of rust caused by *Puccinia malvacearum* on mallow species despite the restricted and scattered nature of host plants. Similarly, in Australia, in years favourable to pathogen development, individual lombardy poplars may be heavily infected with *Melampsora larici-populina* (leaf rust) despite being separated by more than a kilometre from other susceptible hosts. How can the apparent paradox raised by these observations be explained? The answer to this question lies in the relative importance (to disease severity) of infections derived from inoculum produced by the same or other host plants.

Allo- versus *auto-infection*

One measure of the efficiency of inoculum dispersal is the degree to which new infections occurring on a host plant result from inoculum produced by lesions elsewhere on the same plant (auto-infection) or from lesions occurring on other individuals (allo-infection). Because of its dependence on inoculum transmission between plants, the likelihood of allo-infection will be greatly influenced by plant density. Auto-infection, on the other hand, will occur in all plant stands regardless of their density. However, the relative importance of this source of infection will decline with increasing efficiency of inoculum dispersal and host density.

In a typical situation where disease is increasing in a plant population of reasonable size and density, some of the infections occurring on any particular individual will result from auto-infection and some from allo-infection. However, in the case of a totally isolated individual, once a pathogen successfully arrives, disease will increase at a rate determined solely by the level of successful auto-infection. In a simple experiment involving mildew of barley (caused by *Erysiphe graminis hordei*), Burdon & Chilvers (1976*a*) found that the rate of increase of disease on an isolated plant (auto-infection only) was a third of that occurring on individual plants growing in a stand with a density of 31 plants per m^2. Even more rapid rates of increase due solely to auto-infection might occur if inoculum wastage were reduced by the size and shape of the host (for example, a multi-stemmed plant *versus* a single-stemmed one).

Allo- and auto-infection have markedly different effects on the spatial distribution of disease in host plant populations. Because allo-infection involves the spread of disease from one plant to another, through time it will inevitably cause a flattening of disease dispersal gradients. Even in plant populations where individual members are well separated from one another, efficient allo-infection will ensure that disease occurs fairly evenly throughout the population. In contrast, where allo-infection is very limited, disease occurrence will be much more patchy. Once a given plant is infected, the amount of disease on that plant will continue to increase as a result of auto-infection. However, with little or no spread between adjacent hosts, heavily infected plants may occur in close proximity to uninfected ones. Such unevenness of distribution is typical of disease in many natural situations.

This brief consideration of the effects of auto-infection has stressed its interaction with host plant density. Equally, the phenomenon of auto-infection is of direct relevance to the population dynamics of pathogens occurring in mixed host stands. There, the direct consequence of interspersing resistant and susceptible host genotypes is to increase the average distance between susceptible individuals and hence the relative importance of auto-infection.

Mixed plant populations and disease occurrence

Unlike agricultural crops, most natural plant populations are not even-aged, high-density stands of genetically uniform individuals. On the contrary, they are typically mixtures of different species or genotypes of the one species that react differently to a range of environmental factors. Here we are interested in investigating the effects such mixtures have on the size and rate of development of epidemics of plant pathogens. In this we have to rely mainly on information derived from recently developed agricultural systems that deliberately endeavour to increase the level of genetic diversity present in individual crops.

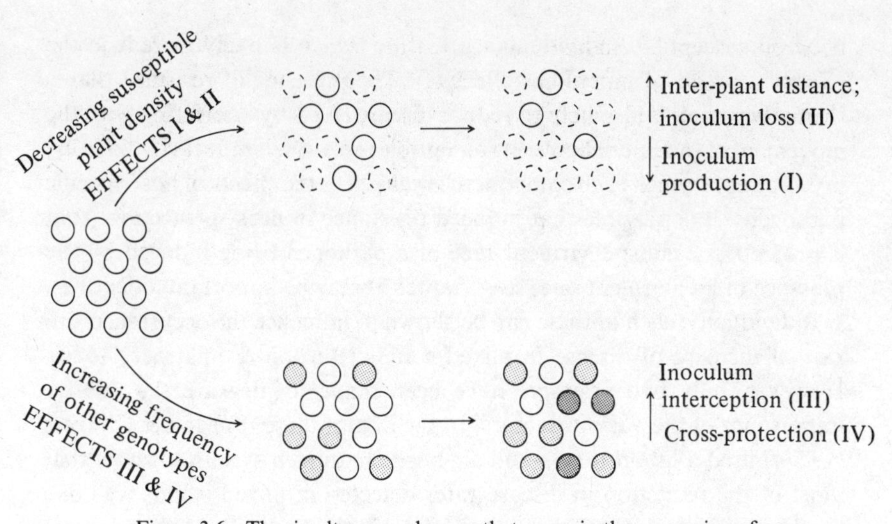

Figure 3.6 The simultaneous changes that occur in the conversion of a pure stand to a mixture of susceptible (unshaded circles) and resistant (shaded circles) species.

How can mixtures reduce disease?
In some mixtures, disease reductions may result from subtle changes in micro-climatic conditions within the mixture affecting the relative success of a pathogen. While such changes may occur in inter-specific mixtures due to radical differences between components in the size, number and disposition of a range of morphological characters, they are far less likely to occur in intra-specific mixtures (for example, varietal mixtures and multilines) where there is little phenotypic difference between component lines. However, such mixtures do show marked reductions in disease levels relative to their component lines (see below). It would seem, therefore, that other mechanisms must exist whereby disease reductions are achieved. The interaction of some of the most obvious of these is illustrated in Figure 3.6.

The replacement of individuals of a uniformly susceptible plant population by resistant plants sets in motion a complex series of interconnected changes which affect the ability of pathogens to survive and reproduce. Some of these changes stem from the actual removal of susceptible plants while others are the consequence of the insertion of resistant plants (Burdon & Chilvers, 1976*b*). The replacement of susceptible plants by resistant ones reduces the amount of susceptible tissue which may become infected. This, in turn, should reduce the amount of inoculum available for subsequent dispersal (effect I, Figure 3.6). The increased distance between susceptible individuals which results from the removal of some of their number increases the average distance that inoculum has to travel between susceptible individuals – a feature which is likely to reduce the effective spread of inoculum (effect II). The presence of resistant plants themselves should also help to reduce disease levels by interfering with the movement of inoculum between susceptible individuals (effect III). Finally, in mixtures in which each component is subject to the effects of host-specific pathogens, it is possible that induced resistance or cross-protection (that is protection against a virulent race of a pathogen being induced by the presence of an avirulent one; see Chapter 5) may be important (effect IV).

Individually, each of these can be shown to influence the occurrence and rate of increase of disease in mixed stands (Burdon & Shattock, 1980). However, only two attempts have been made to measure the relative importance of the various factors. In the first of these, Burdon & Chilvers (1977*a*) used a laboratory-based air-borne pathogen system to show that most of the reduction in disease rates detected in mixed stands was due to the decline in density of susceptible individuals. Only in mixtures containing high frequencies of resistant individuals was there evidence that the interception of inoculum was contributing to disease reduction. In contrast, Chin & Wolfe (1984*a*) used a complex series of two-component mixtures to allow the influences of the reduced density of susceptible plants, the barrier effect of resistant plants and the occurrence of cross-protection to be evaluated separately in a field situation. Overall, they found that in the early stages of plant growth the density effect accounted for most of the observed disease reduction. As plants grew and the disease epidemic developed, however, the barrier and induced protection effects became increasingly important.

The relative importance of each of these mechanisms is likely to vary with the stage of crop growth and epidemic development, host and pathogen genotypes and prevailing environmental conditions (Burdon & Chilvers, 1977*a*; Chin & Wolfe, 1984*a*).

Do mixtures actually reduce disease?

In line with theoretical expectations, reductions in the size of pathogen populations in mixed stands (relative to pure stands) have been observed on numerous occasions in agricultural situations. Reductions have been shown through measurement of several factors: the number of air-borne spores present in or over mixed stands relative to monocultures (Cournoyer, 1970; Browning & McDaniel, cited by Browning, 1974); rates of increase of disease (Leonard, 1969*a*; Berger, 1973; Burdon & Chilvers, 1977*a*; Burdon & Whitbread, 1979); or by assessments of the intensity of disease at a single arbitrary time during crop development. The latter

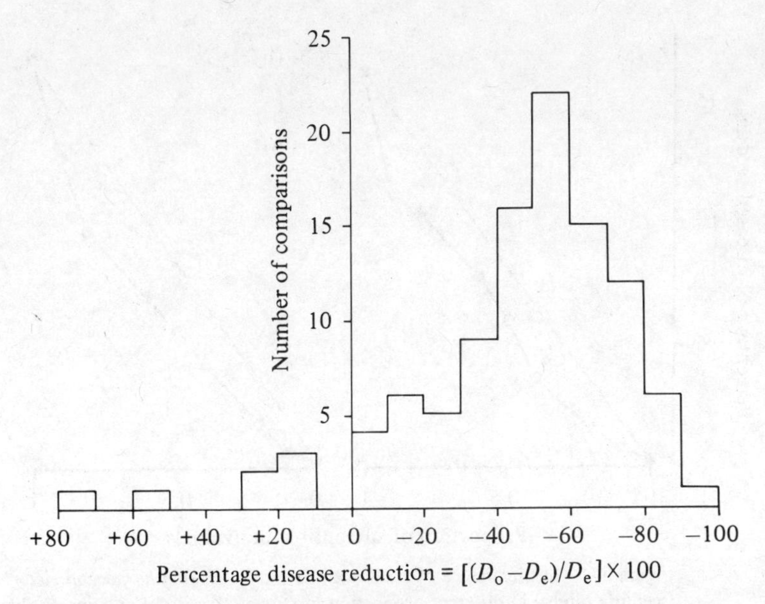

Figure 3.7 Observed level of disease reduction ($D_o - D_e$) in 1:1 two-component mixtures expressed as a percentage of the monocultural means. D_o is the observed disease level and D_e is the mean of that occurring in the monocultures. Negative values represent disease levels less than, and positive values disease values greater than, a simple monocultural mean. (Data derived from Grummer & Roy (1966); Ayanru & Browning (1977); Coyne *et al.*, (1978); Groenewegen & Zadoks (1979); Clark (1980); Hamblin *et al.*, (1980); Day (1981); Fried *et al.*, (1981); Jeger *et al.*, (1981); Wolfe *et al.*, (1981); White (1982).)

approach, by far the simplest, has been used in a large number of experimental investigations where disease levels have been determined in simple, equi-proportioned mixtures. These mixtures may be of different species, different varieties of the one species (varietal mixtures) or different near-isogenic lines of the one variety (multilines). In 103 two-component mixtures representing ten different host–pathogen combinations (Figure 3.7), disease levels were always less than those recorded in the most affected monoculture. In fact, in most cases disease levels were often very substantially below the arithmetic mean of those recorded in pure stands of the mixture components.

In one or two cases, disease levels have been investigated more extensively by examining the occurrence of disease in a series of mixtures of different proportions. One of these studies investigated the effects of seven different mixtures of two oat lines, one resistant and the other susceptible to *Helminthosporium victoriae* (seedling blight), on the level of disease

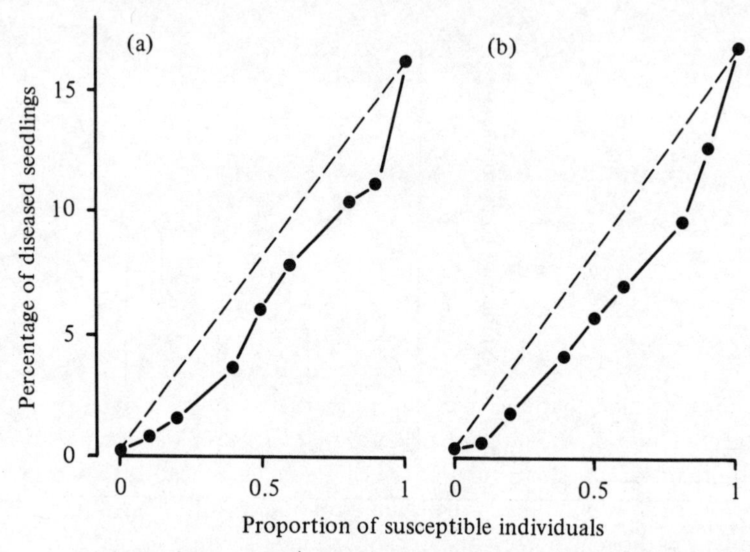

Figure 3.8 The incidence of seedlings affected by *Helminthosporium victoriae* (seedling blight) in different proportion mixtures of resistant and susceptible oat lines. (a) and (b) are two experiments without and with an undersown crop of lucerne. The dashed line represents the expected percentage of diseased seedlings in the appropriate mixture. (Derived from data of Ayanru & Browning, 1977.)

caused by this pathogen (Ayanru & Browning, 1977). Again, in all mixtures, disease levels were less than those predicted by simple proportionality (Figure 3.8). Moreover, even in mixtures composed mainly of susceptible individuals ($\geqslant 80\%$), disease levels were markedly reduced.

There are, however, a few cases where higher levels have been recorded in mixtures. One of these, die-back of *Dactylis glomerata* caused by *Rhizoctonia solani* (Chamblee, 1958), possibly reflects a stimulation of pathogen growth by exudates from the roots of resistant individuals. A more intriguing example occurred in a mixture of the barley cultivars 'Tystofte Prentice' and 'Svalof Freja' where a sterility interaction was found (Sandfaer, 1968). In 1:1 mixtures the percentage of sterile flowers of 'Svalof Freja' increased six-fold whereas the percentage of sterile flowers of 'Tystofte Prentice' fell by one fifth. Thorough testing showed this phenomenon to be caused by barley stripe mosaic virus carried by 'Tystofte Prentice' (Sandfaer, 1970).

The effects of these reductions in the occurrence of disease on the reproductive capacity of individual members of such host populations are considered in Chapter 6.

4

The genetic basis of disease resistance in plants

One truism of plant pathology is that most plants are resistant to most pathogens. Thus potatoes are not attacked by *Puccinia graminis tritici*, the causal organism of stem rust of wheat, nor is wheat attacked by *Phytophthora infestans*, the causal organism of potato late blight. Although some plant pathogens possess remarkably wide host ranges (for example, *Pythium* and *Rhizoctonia* species responsible for seedling damping-off diseases), the majority tend to be restricted to a few closely related host species. Even within individual host species, however, casual observation and rigorous experimental investigation have often found some resistant individuals within otherwise apparently uniformly susceptible species. The aim of this chapter is to develop an understanding of the extent and genetic basis of such disease resistance mechanisms in plants.

The expression of disease resistance

A wide variety of disease resistance mechanisms has been found in plants. Recognition of the existence and epidemiological consequences of these forms of resistance is of considerable importance in developing an understanding of their likely role in plant populations. In general, resistance mechanisms can be divided into two broad categories that reflect the degree of interaction occurring between host and pathogen. Active resistance mechanisms occur as a result of continuing interactive responses. Passive resistance, on the other hand, encompasses mechanisms resulting from past interactions between host and pathogen which have led, through selection, to an avoidance of contact or mechanisms whereby the detrimental effects of disease are at least partly absorbed by the infected plants (tolerance).

Active disease resistance mechanisms

Disease resistance terminology

Although all active disease resistance mechanisms may be broadly attributable to the outcome of a continuing active response of a host to an invading pathogen, the actual phenotypic expression of these mechanisms is very varied. This diversity and the essentially continuously variable nature of disease resistance in plants has produced enormous problems in terminology. Active disease resistance mechanisms are not easily subdivided into unambiguous categories but for a variety of reasons it has often been convenient to attempt to do so. Unfortunately, however, the range of terms used to describe such sub-categories is large (e.g. horizontal, vertical, race specific, race non-specific, oligogenic, polygenic, multigenic, seedling, adult plant, qualitative, quantitative, partial, uniform, generalized, rate-reducing, slow-rusting, slow-mildewing, dilatory, discriminatory, durable and field resistance). Moreover, not only do many of these terms have very similar meanings, but at various times the same term has been imbued with a variety of different meanings. Confusion has inevitably followed.

Originally, van der Plank (1963) in his influential work on epidemiology divided active resistance into horizontal and vertical components. Horizontal resistance was defined as that resistance effective against all races of a pathogen and vertical resistance as that effective against some, but not other, races of the pathogen. However, these genetically based definitions were confounded with an inaccurate assessment of their epidemiological consequences. Vertical resistance was seen only to delay the onset of an epidemic while horizontal resistance reduced the rate at which an epidemic developed. As Browning (1979) has pointed out, subsequent attempts at redefinition while retaining the same terms (Robinson, 1976; Nelson, 1978; van der Plank, 1978) have only led to further confusion and render these terms suspect.

Because of this problem Browning and his colleagues (Browning, Simons & Torres, 1977; Browning, 1979) have attempted to classify active resistance mechanisms entirely according to their epidemiological consequences. Resistance which hampers the rate of pathogen development and hence reduces pustule size, propagule production and/or longevity, affects the rate of increase of an epidemic and has been called dilatory resistance. On the other hand, resistance that selectively screens out some pathogen genotypes and not others reduces the initial amount of inoculum and hence only delays the epidemic in time (see Chapter 3). This was designated discriminatory resistance. Unfortunately such a division is also somewhat confusing as resistance characterized by infection types with even only very sparsely sporulating lesions must be classed as dilatory although a discriminatory effect occurs. Moreover, while this distinction may have some merit in those agricultural systems where all host plants react identically, in deliberately diversified crop mixtures and in many wild plant–pathogen interactions the epidemiological effects of differences between these two effects are blurred even further. There the presence of a range of host types each discriminating against different pathogen biotypes produces a continuing (rather than a once-only) reduction in the amount of effective inoculum. This will cause a reduction in the rate of development of an epidemic.

An alternative approach and the one used in this book basically adopts van der Plank's (1963) genetic division of resistance but makes no *a priori* claims concerning the epidemiological effects of either category. Resistance that is specific to a particular virulence character (Scott *et al.*, 1979) and is recognized by a differential interaction between host and pathogen genotypes is defined as *race specific*. In contrast, *race non-specific resistance* is defined as resistance that is not specific to particular virulence characters. It is implied in an absence of adaptive matching of host and pathogen genotypes, so that an increase in race non-specific resistance results in less successful parasitism by all pathogen individuals regardless of their genotype (Scott *et al.*, 1979). (Because race non-specificity cannot be proved absolutely, this type of resistance should strictly be referred to as *apparent* race non-specific resistance.) The division between race specific and race non-specific resistance often has the added practical advantage of separating forms of resistance which are determined relatively simply at a single time from those that may require considerably more effort to characterize. The expression and detection of these two categories of active resistance are now considered. However, it must always be remembered that both systems co-exist in nature, where they merge in a continuum of genetic effects that range from one extreme to the other (Day, 1974).

Race specific resistance

The most highly developed system for recognizing and categorizing differences in the resistance of host plants measures race specific resistance. In this system the degree of resistance or susceptibility of a host to various different pathogen races is expressed phenotypically as differences in the visual appearance of lesions. In fully susceptible hosts, a completely compatible interaction between host and pathogen results in an infection type characterized by large lesions or pustules surrounded

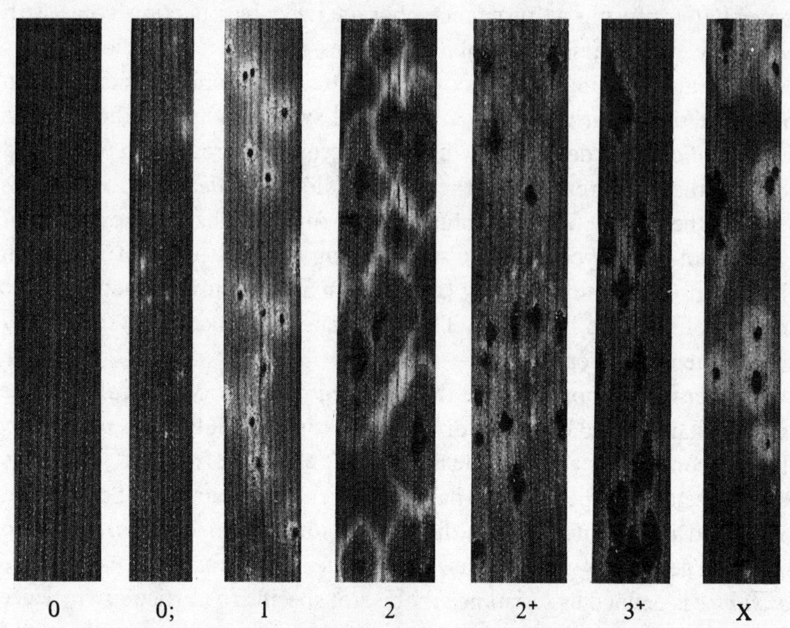

Figure 4.1 Variation in infection types that may be produced by races of *Puccinia graminis tritici* (stem rust) on differential varieties of *Triticum* spp. The infection types are scored according to the standard 0–4 scale. (0, immune, no visible signs of infection; 0;, hypersensitive flecks, no uredia; 1, minute uredia surrounded by distinct necrotic areas; 2, uredia small, usually in green islands surrounded by chlorotic or necrotic borders; 2^+, uredia medium in size, less necrosis; 3^+, uredia large and often coalescing, limited chlorosis; X, heterogeneous reactions, uredia variable in size.) (Photographic plate courtesy of Dr R. A. McIntosh and Mr D. J. S. Gow, University of Sydney, Plant Breeding Institute, Castle Hill.)

by apparently healthy tissue. Chlorotic or necrotic regions are absent. On the other hand, a range of different resistance reactions may occur. These are identified by varying degrees of disease development and pathogen sporulation. Depending upon the particular combination of host and pathogen isolate, a resistant response may appear as a complete lack of macroscopically visible symptoms, as non-sporulating lesions or flecks (a hypersensitive response), or as the presence of pustules of varying fecundity surrounded by regions of chlorotic or necrotic tissue.

In studies of the interaction between cereal rust pathogens and their economically important hosts, the classification of these differences has been standardized into a series of separate infection types using a semi-continuous 0 to 4 scale (Stakman & Harrar, 1957; Simons, 1970). On this scale, complete susceptibility is designated by infection type '4' and complete resistance or immunity to infection by type '0'. Between these two extremes lies a continuum of infection type responses (Figure 4.1). The degree to which these can be separated depends upon the reproducibility of individual classifications. Although up to 30 classificatory steps have been used in applying this scale to the interaction between *Puccinia graminis tritici* (stem rust) and *Triticum* sp. (Luig, 1983), such a fine level of distinction is neither necessary nor likely to be possible in the great majority of ecological studies. Similar classificatory scales have been developed for other plant–pathogen interactions (for example, *Erysiphe graminis* (powdery mildew) on *Triticum* spp.).

Infection type responses are typically determined on seedling plants and the responses thus detected are usually manifest throughout the life of the plant. Occasionally, however, plants that are susceptible at the seedling stage are resistant as adults ('adult plant resistance') or, conversely, resistance expressed at the seedling stage remains effective only while the plant is young.

The classification of individual host plants as susceptible or resistant according to their infection type response is the main means of identifying race specific resistance. It is resistance of this type, identified in this way, that has been most widely used in protecting cereals and a wide variety of other agricultural crops against pathogens. Because it may result in resistant reactions that produce either no spores or only a restricted number, this type of resistance can affect both the initial level of effective inoculum and the rate of increase of a disease epidemic.

Comparable differences between resistant and susceptible infection types occur in all host–pathogen interactions. The precise nature and phenotypic expression of these differences will be determined by the aetiology of the particular pathogen involved. However, the ways in which these differences may be used to investigate the genetics of host–pathogen interactions is the same for all such combinations.

Race non-specific resistance

Race non-specific resistance is presented uniformly against all races of a pathogen. Its presence is independent of the occurrence of race specific resistance, although it is often easiest to document in host genotypes that exhibit fully susceptible infection type responses. In contrast to race specific resistance, race non-specific resistance is relative in nature. It can only be detected by comparing the disease reaction occurring on different host genotypes. Moreover, it may be expressed in a variety of different ways. These are perhaps best illustrated by considering the fate of a

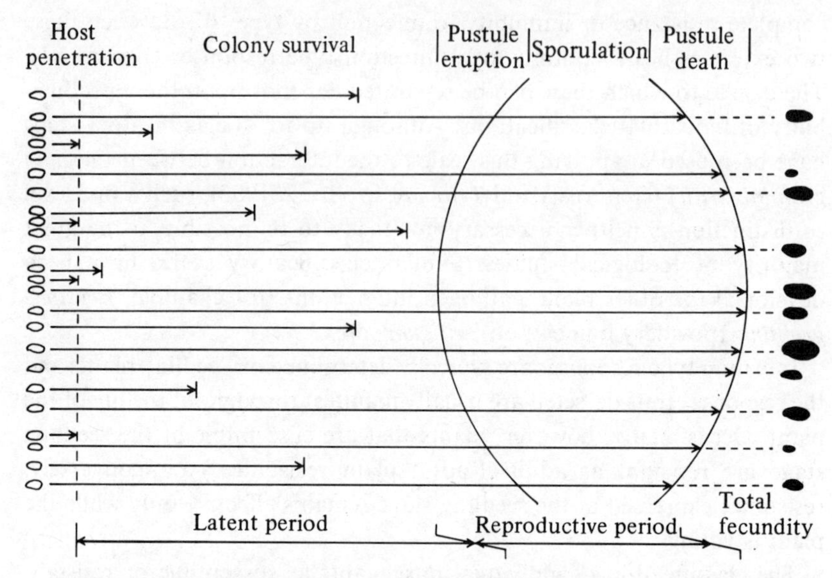

Figure 4.2 Aspects of race non-specific resistance showing, diagrammatically, the various stages at which this type of resistance operates in host–pathogen interactions.

population of spores from the time of its arrival on a host plant through to its final demise (Figure 4.2). Reductions in the frequency of successful host penetration and of the survival of developing fungal colonies will result in a reduction in the actual number of productive lesions; restrictions in the amount of hyphal growth may affect pustule size and the rate and duration of spore production; finally, increases in the length of the latent period will increase the average generation time. All of these features are expressions of race non-specific resistance that are likely to have profound epidemiological consequences without affecting the phenotypic expression of disease symptoms.

In a detailed study of the relative ability of isolates of *Puccinia hordei* (leaf rust) to infect a range of barley genotypes, Niks (1982) found that the proportion of successful and failed infections resulting from a uniform inoculation differed significantly between host genotypes (Table 4.1). Microscopic examination of leaves showed these differences to be due mainly to variations in the incidence of colony abortion. However, a non-significant change in the frequency of successful host penetration was also observed. Statistically significant differences in the mean number of pustules occurring per leaf have been observed in oat cultivars and breeding lines infected with *Puccinia coronata* (crown rust) (Politowski &

Table 4.1 *The proportion of successful and aborted infections of* Puccinia hordei *(leaf rust) occurring on seven different genotypes of barley. Two separate experiments (Series I and II) were carried out using the barley cultivar 'Vada' as a consistent control (data from Niks, 1982)*

Series	Barley genotype	Proportion of infections successful	Proportion of infections failing due to: Non-penetration	Early abortion	Late abortion
I	Vada	0.40[a] [a]	0.03	0.54[a]	0.08
	C-29	0.72[b]	0.03	0.17[b,c]	0.09
	C-70	0.74[b]	0.06	0.15[c]	0.07
	C-92	0.53[a]	0.05	0.35[b]	0.14
II	Vada	0.38[a]	0.03	0.59[a]	0.03
	C-120	0.49[a]	0.06	0.17[b]	0.36
	C-123	0.42[a]	0.05	0.56[a]	0.00
	C-197	0.81[b]	0.04	0.12[b]	0.02

[a]Different letters (a, b, c) in each experimental series indicate that values differ significantly at $P \leqslant 0.01$ (Duncan's multiple range test).

Browning, 1978). However, in this case it should be noted that although several varieties showed the same general pattern of response to two different races of this pathogen, in the variety 'Red Rustproof' there was a differential effect – an apparent case of race specificity! The actual causes of the difference in resistance of the oat varieties were not determined but may again reflect variations in the frequency of colony death or of spore germination and host penetration.

Significant variation in the frequency of host penetration has been recorded on a number of occasions. In an assessment of the response of 17 wheat varieties to *Puccinia graminis tritici*, the frequency of successful infection events ranged from 0.003 to 0.193 (Brown & Shipton, 1964). Even when comparisons are restricted to varieties showing identical infection types the magnitude of differences was almost as great (0.003 to 0.172). A variety of morphological and physiological features have been implicated in resistance of this kind. A discussion of some of these with respect to rust fungi is given by Hooker (1967).

Differences in the length of the latent period, due presumably to restrictions on the rate of colony development, have been suggested as an additional means whereby resistance may be expressed (van der Plank, 1968; Nelson, 1973). Such differences have been detected between geno-

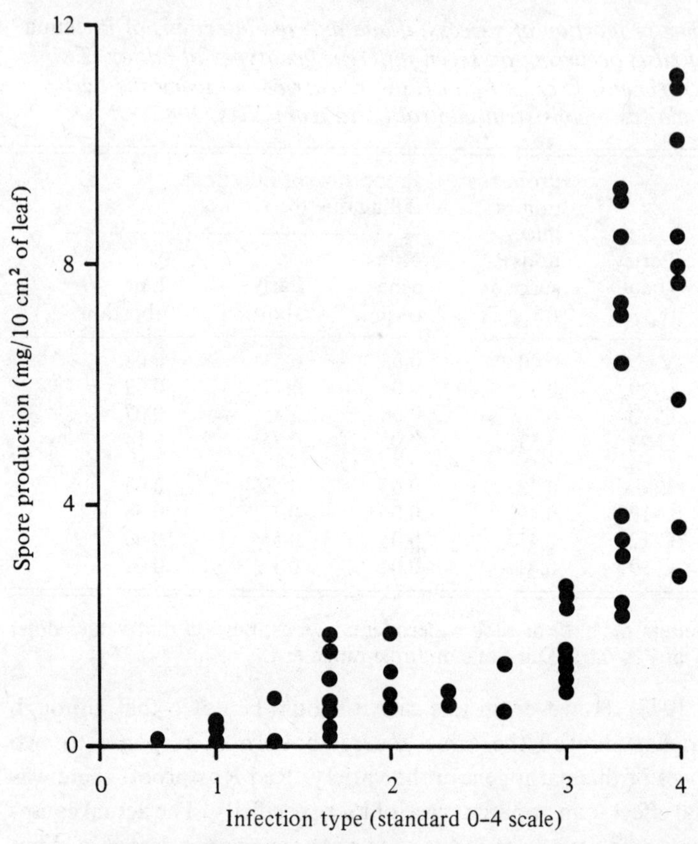

Figure 4.3 Spore production (mg/10 cm² of leaf) from single wheat seedlings infected with *Puccinia striiformis* (stripe rust) plotted against infection type on the standard 0–4 scale. (Redrawn from Johnson & Taylor (1976) and reproduced, with permission, from the *Annual Review of Phytopathology*, Vol. **14**. © 1976 by Annual Reviews Inc.)

types of a range of species, including wheat infected with *Puccinia recondita tritici* (leaf rust) (Ohm & Shaner, 1976), rye infected with *P. recondita recondita* (Parlevliet, 1977) and members of the genus *Glycine* infected with *Phakopsora pachyrhizi* (rust). In the latter example, accessions of four native species collected from various parts of eastern Australia showed marked intra-specific differences in the length of the latent period of this pathogen. In *G. canescens* the latent periods of 20 accessions ranged from 9 to 24 days while in *G. clandestina* those of 34 accessions ranged from 9 to 17 days (Burdon & Marshall, 1981*a*). Similarly, the prolonged latent period of *Puccinia hordei* on the barley cultivar 'Vada' relative to other genotypes is well known (Clifford, 1972; Parlevliet & van Ommeron, 1975; Parlevliet, 1979).

Finally, race non-specific resistance may also be expressed in a form which directly affects the fecundity of sporulating lesions. This can occur through subtle variations in pustule size not readily detected in infection type ratings, through variation in the rate of spore production or through differences in the productive life of individual pustules. Johnson & Taylor (1972, 1976) have found that slight changes in infection types produced by *Puccinia striiformis* (stripe rust) on wheat seedlings can be associated with relatively large differences in spore production (Figure 4.3). This was particularly the case when relatively susceptible infection type reactions were compared. In one experiment, plants with identical, fully susceptible infection types showed a four-fold variation in spore production per unit area of leaf.

In the field such responses may affect the rate of disease development. In an analysis of the response of the barley varieties 'Knox' and 'Vermillion' to infection by *Erysiphe graminis*, Shaner (1973) found that the combined effects of reduced size and productivity of fungal colonies resulted in lower levels of disease on 'Knox' than on 'Vermillion'.

The epidemiological effect of those expressions of race non-specific resistance which lengthen the latent period, shorten the infectious period or reduce the productivity of individual lesions or pustules is clearly to reduce the rate of disease increase. However, race non-specific resistance that is expressed as reductions in the frequency of successful host penetration and colony establishment may also affect the level of effective inoculum.

Environmental and ontogenetic effects on resistance

The phenotypic expression of all active disease resistance mechanisms may be markedly affected both by variations in environmental conditions and by ontogenetic changes in the host plant. As we shall see in Chapter 5, changes in light intensity, temperature and the nutrient status of host plants all affect interactions occurring between host and pathogen during the development of an infection. For many pathogens, for example those responsible for rust and mildew diseases, poor or unbalanced nutrient regimes may result in longer latent periods, smaller pustules, poorer sporulation and shorter reproductive periods than normally encountered under optimal nutrient conditions. Similarly, increases in temperature often result in shorter latent periods and generally accelerated pathogen development (Rowell, 1984). In a few celebrated cases, for example the

Lr20 and *Sr15* genes in wheat, temperature differences of only a few degrees may produce dramatic changes in the phenotypic expression of disease (see Figure 5.3).

Of equal importance to the effects of environmental variation on the expression of resistance are the effects of ontogenetic changes that occur as host plants grow and mature. Although plants usually become more resistant with age, examples of three other changes in resistance with increasing ontogenetic age have been recognized (see Chapter 5). However, for pathogens responsible for rust diseases, resistant reactions characterized by low (resistant) infection types are generally expressed throughout the life of the plant. In a few instances, though, increased susceptibility has been associated with ageing (Duff, 1954) or specific growth stages (Samborski & Ostapyk, 1959).

Passive disease resistance mechanisms

Disease tolerance

Disease tolerance is a measure of the relative ranking of the performance of two or more lines of a host species in the presence of disease expressed as a fraction of their respective performances in its absence. Phenotypically, tolerant susceptible individuals appear heavily diseased without sustaining severe yield losses, while similarly infected non-tolerant individuals show marked reductions in yield. However, because the recognition of tolerance is based on final yields, its meaning has become confused and its measurement confounded by the potential differential presence of race non-specific resistance mechanisms.

The accurate assessment of tolerance requires either an equality of response of different host lines to a pathogen or that such differences that occur are recognized and accounted for during subsequent analysis. Even minor phenotypic differences in race non-specific resistance mechanisms, when compounded during the course of a single epidemic, over many pathogen generations, may produce major epidemiological effects. As Schafer (1971) and Gaunt (1981) have pointed out, careful measurement is required to distinguish between the effects of such race non-specific resistance and true tolerance. Not only must the various host lines under consideration be equally susceptible but this similarity must generally be carried through to the occurrence of equivalent levels of disease at all stages of the disease epidemic. Measures of disease severity at a single time (usually near harvest) will often fail to detect major differences in disease levels that occurred earlier in the growing season. Even in situations where the cumulative level of disease over time is equivalent, fluctuations in the vulnerability of hosts to disease damage at different stages in the growth cycle may affect final yields materially. The problem produced by the interacting effect of variations in the duration, timing and severity of disease on final plant productivity is considered in detail in Chapter 2.

In practice, the need for strict equivalence of disease levels is not always necessary. In situations where one variety supports significantly more disease than another while suffering the same or a lesser relative yield loss, the former variety can be classified as the more tolerant. However, if the yield loss incurred by the more severely afflicted line is greater than the less diseased one, a lack of linearity in the relation between disease severity and yield loss precludes the determination of which, if either, line is the more tolerant (Schafer, 1971).

The difficulties of maintaining accurately uniform levels of disease severity and the complicating effects of undetected resistance have frequently resulted in the misclassification of cereal crop varieties as tolerant when, in reality, they owe their relatively good performance in the presence of disease to apparently insignificant levels of race non-specific resistance. The 'tolerance' of the oat variety Benton to crown rust caused by *Puccinia coronata* (Caldwell *et al.*, 1958) was subsequently attributed to a slower rate of development of the pathogen on this cultivar (Clifford, 1968). Similarly, the relatively high yields of various other lines of oats and wheat in the presence of disease are now known to result from previously undetected race non-specific resistance (Hayden, 1956; Michel & Simons, 1971). Despite these problems, careful experimentation, incorporating measures designed to differentiate between tolerance and active disease resistance mechanisms, has shown that tolerance does exist (Politowski & Browning, 1978). However, these workers suggest that, in agriculturally important cereals at least, true tolerance is a rare phenomenon.

In the light of these findings it is difficult to assess the importance of true tolerance in non-agricultural plant populations. The considerable technical difficulties associated with the accurate measurement of tolerance in genetically relatively uniform agricultural crops are greatly compounded in wild plant populations. In those populations, it would be necessary to ensure that a wide range of life history traits were relatively unaffected before a claim of true tolerance could be sustained. As a consequence, the separation of true tolerance and minor race non-specific resistance effects is unlikely to be achieved for some time. However, we cannot assume that the frequency and importance of tolerance in non-agricultural plant populations will necessarily be reflected by its role in agricultural ones. During the early stages of plant improvement programmes, phenotypically susceptible lines with genes that help maintain performance under conditions of high disease severity are likely to be discarded along with non-tolerant susceptible lines (Clark, 1966). As a result, at present all we can say is that tolerance is likely to occur and may play a more important role in non-agricultural plant populations than in agricultural crops.

Disease avoidance

Disease avoidance and disease escape are two frequently confused terms which cover different aspects of the failure of host plants to become infected by their pathogens. In both cases, hosts are susceptible but infection and subsequent disease development is prevented through the presence of temporal, spatial or physical barriers. However, an essential distinction between these two terms lies in their bases. Disease avoidance has a genetic basis while disease escape is the result of a fortuitous set of circumstances and should not be seen as a resistance mechanism. Thus disease escape is seen here as a lack of infection resulting from the failure of host and pathogen to come into physical contact although both are present in the same environment at the same time. It is a simple spatial phenomenon, instances of which are often seen in the early stages of disease epidemics, particularly where the initial inoculum is carried into a host population by wind or insect vectors. The extreme patchiness of viral infections of many crops (see Gibbs & Harrison, 1976) or the development of fungal epidemics from distinct foci provide examples of such disease escape.

Disease avoidance, on the other hand, encompasses mechanisms under the genetic control of the host which may have been selectively favoured by past pathogen-induced disease pressure on the relevant host population. Mechanisms of disease avoidance are varied and are particularly included here to stress the possible epidemiological consequences of a wide range of morphological and physiological host plant characters. From a crop breeding point of view these are not traditionally viewed as disease resistance mechanisms. However, from an ecological stance, they may be of great importance in contributing to the survival of individual hosts.

The most obvious physiological characters involved in the avoidance of disease by plants are those which affect a temporal separation of susceptible stages in the life cycle of the host and physical environmental conditions particularly favourable for the rapid growth and development of the pathogen. Precocious germination, rapid 'hardening-off' of seedling tissues (Agrios, 1980) or an early flowering habit may all minimize contact between host and pathogen during times of extreme host vulnerability. Burdon (1982) has suggested that earliness may be important in reducing the effects of crown rust on populations of *Avena barbata* growing in New South Wales, Australia. In other host–pathogen interactions, however, late flowering may be just as advantageous. Apparently, although wheat that matures early in Israel avoids significant damage from rusts, late-maturing varieties avoid the detrimental effects of *Septoria* infections (Dinoor & Eshed, 1984).

Various morphological attributes may also result in the successful avoidance of disease by otherwise susceptible host plants. Classical examples of this type of resistance are found in the physical protection afforded by the closed-flower habit of many cereals to infection by *Claviceps purpurea* (ergot) and *Ustilago nuda* (loose smut). Both of these pathogens infect flowers and gain entry to the ovule through the stigma. They do this at the same time as these flowers are receptive to pollination. Varieties of wheat and barley in which florets remain closed, at least until pollination has occurred, very rarely suffer natural infections. However, infections can easily be produced in these varieties if stigmas are inoculated experimentally at anthesis (Russell, 1978).

Selective forces other than pathogen pressure (for example, seasonal drought) may have been primarily responsible for changes in such characters as flowering time. However, in many cases the action of these agents have almost certainly been reinforced by pathogen-induced reductions in the reproductive output of host plants. In many non-agricultural plant communities features like early flowering are common. Just how important these are as a means of protection for individual plants is very difficult to assess but certainly they should not be totally neglected.

The genetic basis of disease resistance

Studies of disease resistance in plants have tended to concentrate on determining the genetic basis of obvious differences in infection types. Such differences, whether qualitative or quantitative, are generally easily assigned to a few categories, thus simplifying the final interpretation of data. In a review of almost 800 papers covering the inheritance of resistance in plants to a range of biotrophic fungal pathogens, Person & Sidhu (1971) and Sidhu (1975) found that more than 90% of studies concerned resistance characterized by marked phenotypic differences between resistant and susceptible infection type responses (race specific resistance). Less than 10% dealt with phenotypic differences that were so minor that consistent discrete classification was not possible.

Race specific resistance

Where care has been taken to ensure the purity of pathogen races, the majority of studies of race specific resistance have shown that resistance is inherited according to simple Mendelian principles (Day, 1974; Ellingboe, 1976). In general, resistance is completely dominant to susceptibility so that in an F_2 population segregating for a single resistance gene, a 3:1 ratio of phenotypically resistant to susceptible individuals is observed. Only about 10% of resistance genes are completely recessive in their expression. However, these are to be found in a wide range of host–pathogen combinations (Sidhu, 1975), including the resistance produced by some genes to *Puccinia graminis tritici* (stem rust) and *P. striiformis* (stripe rust) in wheat (Lupton & Macer, 1962; Berg, Gough & Williams, 1963), to *Erysiphe graminis* (mildew) in peas (Heringa, Van Norel & Tazelaar, 1969) and to *Puccinia sorghi* (rust) in maize (Malm & Hooker, 1962). Incomplete dominance, expressed in individuals heterozygous for particular resistance genes as an intermediate disease response, is poorly documented although it undoubtedly exists in a number of host–pathogen systems.

Genetic analyses have often revealed the occurrence of two or more resistance genes within a single host genotype. This is particularly the case in agriculturally important cereal varieties where one deliberate disease control strategy has been to incorporate many genes for resistance within single varieties. Modern Australian wheat varieties like Gamut and Timgalen carry, respectively, four and five known genes for resistance to races of *Puccinia graminis tritici*. In a few cases where detailed studies have been carried out, multiple genes for resistance have also been found in wild plant populations (see Chapter 7). When two or more resistance genes occur within the one plant they may function independently of one another or may interact (Day, 1974). Commonly, genes that condition higher levels of resistance to pathogens that cause rust diseases are epistatic to those that condition lower levels of resistance. An illustration is given by the two resistance genes *Sr5* and *Sr9b* that confer resistance to *P. graminis tritici* in wheat. When challenged by an avirulent pathogen race, the infection type of seedlings homozygous for *Sr5* is characterized by an immune or hypersensitive fleck reaction and that of seedlings homozygous for *Sr9b* by small chlorotic pustules (Luig & Rajaram, 1972). When these genes are

Table 4.2 *Infection type responses*[a] *of homozygous lines of wheat carrying particular combinations of resistance genes when challenged with an avirulent race (race 24–4, 5) of* Puccinia graminis tritici *(stem rust) (data from Luig & Rajaram, 1972)*

Host genotypes	Temperature (±1 °C)			
	15	18	21	24
Sr5Sr5	0;	0	0;	0;
Sr9bSr9b	2,3,c	2^-,3,c	2^-,3,c	3^-,c
Sr13Sr13	2	2^-	$2^=$,2	$2^=$,2
Sr5Sr5Sr9bSr9b	0	0	0	0
Sr5Sr5Sr13Sr13	0	0	0	0

[a]Infection types according to the standard Stakman scale.
0 = immune; 0; = some hypersensitive flecks; 2 = moderately resistant, medium-sized uredia with chlorotic and necrotic halos; 3 = susceptible, large pustules surrounded by chlorotic ring; c = chlorosis; $^-$ and $^=$ indicate infection types slightly less than the noted values; two or more values indicate a range of infection types. See Figure 5.3 for illustration of Stakman scale.

present together in the one host genotype, the challenge of a pathogen genotype avirulent to both induces a hypersensitive fleck reaction (Table 4.2). The phenotypic expression of the resistance gene conditioning small chlorotic pustules is completely hidden by that conditioning a hypersensitive fleck. In Table 4.2 a further example involving the genes *Sr5* and *Sr13* is also given.

Multiple allelism is another phenomenon that occurs among resistance genes in a number of biotrophic fungal pathogens (Sidhu, 1975). Without doubt the best known and most extensive of these is that found in the genes in *Linum usitatissimum* that confer resistance to *Melampsora lini* (flax rust). There, 29 resistance genes are grouped into five allelic series (Lawrence, Mayo & Shepherd, 1981). At the other extreme, in wheat only two loci with multiple alleles are known amongst 36 genes for resistance to *Puccinia graminis tritici* (Luig, 1983). These loci incorporate two and six alleles, respectively.

Other forms of gene interaction have been documented with respect to the expression of resistance. These include: complementary gene interactions, where resistance is dependent on the presence of two or more genes; modifier genes which either enhance or diminish the expression of resistance genes; and reversals of dominance resulting from a change in the genetic background against which a gene is expressed. The intricacies of such interactions are not important in the context of this book. Detailed assessments of their effects are to be found elsewhere (Day, 1974). Aberrant or epistatic ratios may also result from the interaction of two or more pathogens. Sidhu & Webster (1977) and Sidhu (1981) have discussed this problem in detail and shown how the sequential attack of *Fusarium oxysporum lycopersici* (wilt) and *Verticillium albo-atrum* (blight) on segregating F_2 families of tomato may convert a classic dihybrid 9:3:3:1 ratio into a 12:3:1 ratio.

Race non-specific resistance

In comparison to race specific resistance relatively little is known about the precise genetic control of race non-specific resistance. Generally, however, it appears to be determined by many genes, each with a small phenotypic effect (Day, 1974; Parlevliet, 1979). Breeding experiments involving resistance of this type have typically found F_2 segregating populations to display responses ranging continuously between, and sometimes even beyond (Hooker, 1967), those of the parental genotypes. (This is in direct contrast to race specific resistance which can usually be divided into a few discrete categories.)

Selection for race non-specific resistance can be carried out without the need for determination of the precise control involved by repeated culling of susceptible individuals in successive generations. Using this approach Lupton & Johnson (1970) concluded that the resistance possessed by the wheat variety 'Little Joss' to *Puccinia striiformis* (stripe rust) was complex. Even in an F_5 generation originating from a cross between this variety and the highly susceptible 'Nord-Desprez', a continuous, though skewed, distribution of disease reactions was observed. Jones, O'Reilly & Davies (1983) tried to determine the actual number of genes involved in the race non-specific resistance of the oat variety 'Maldwyn' to *Erysiphe graminis* (powdery mildew). Resistance to this variety is mainly expressed as a reduction in primary penetration and haustorial efficiency (Carver & Carr, 1978). Using an F_3 segregating population these workers estimated the minimum number of effective resistance factors occurring in 'Maldwyn' to be between four and nine. No dominance was apparent and the genes functioned additively. The determination of the number and mode of action of genes involved in race non-specific resistance is clearly much more complex than the determination of those involved in race specific resistance.

The genetic basis of pathogenicity

Compared with investigations of disease resistance, studies on the inheritance of pathogenicity are limited (Person & Sidhu, 1971; Sidhu, 1975, 1980). However, a series of detailed investigations has been carried out on the interaction between *Linum usitatissimum* and *Melampsora lini* (see Flor, 1956, 1971; Lawrence *et al.*, 1981). Less intensive studies have been made on a range of other pathogens, including those responsible for rust and smut diseases of cereals. All these studies have shown that avirulence is usually dominant to virulence, that these characters are monogenically based and inherited in a normal Mendelian manner, and that allelism is extremely rare (Day, 1974; Ellingboe, 1976; Sidhu, 1980). As with the genetic basis of disease resistance it is not within the scope of this book to deal with the finer details and complexities of the genetic basis of pathogenicity. These have been considered in detail elsewhere (e.g. Day, 1974).

The gene-for-gene hypothesis

In genetic studies of race specific resistance, segregating populations of either host or pathogen are scanned for visual differences in infection type. Individuals may then be assigned to one of a restricted number of phenotypic categories. In this way the genetic basis of resistance and virulence has been identified in many host–pathogen combinations. By far the most complete of such analyses has been achieved in the interaction between *Linum usitatissimum* and *Melampsora lini*. Although numerous workers have contributed to our knowledge of this system, all the early pioneering work was carried out by H. H. Flor who identified 26 factors in the host and pathogen for resistance and virulence, respectively (see Flor, 1956, 1971). On varieties of *Linum* carrying single resistance genes, segregating F_2 families of the pathogen gave monofactorial ratios of avirulent and virulent individuals. Bi- or tri-factorial ratios occurred on host varieties carrying two or three resistance genes, respectively (Flor, 1971). Similar ratios were obtained when F_2 families of the host segregating for one, two or three resistance genes were challenged with single pathogen genotypes.

The genetic principles uncovered in these experiments led Flor to propose a gene-for-gene hypothesis in which he envisaged that 'for each gene determining resistance in the host there is a corresponding gene in the parasite with which it specifically interacts' (Flor, 1951). In this model the occurrence of a resistant reaction is dependent on both the presence of genes for resistance in the host and the corresponding genes for

Table 4.3 *Genetic interactions in the gene-for-gene model involving one gene in both host and a diploid pathogen.* (R *and* r, *alleles for resistance and susceptibility;* V *and* v, *corresponding alleles for avirulence and virulence;* −, *resistant reaction;* +, *susceptible reaction*)

(*a*) *Complete set of host and pathogen interactions*

Pathogen genotype	Host genotype *RR*	*Rr*	*rr*
VV	−	−	+
Vv	−	−	+
vv	+	+	+

(*b*) *Quadratic check*

Pathogen genotype	Host genotype *R–*	*rr*
V–	−	+
vv	+	+

avirulence in the pathogen. In a diploid host–pathogen interaction, the characteristic pattern of such interactions results in nine genetically unique combinations of host and pathogen genotypes (Table 4.3*a*). However, because expression of the dominant allele is nearly always complete, only two phenotypic expressions (resistant, susceptible) are usually observed. As a result, this pattern can be simplified into the 'quadratic check' by combining together the reaction types of the dominant host and pathogen phenotypes (Table 4.3*b*). The patterns of interaction shown in Table 4.3 occur only because genes for resistance and avirulence are assumed to interact specifically to produce a resistant reaction. When this occurs the simplified pattern becomes independent of dominance relationships (Person & Mayo, 1974).

In a two-gene model in which two different resistance genes are represented by R1 and R2, and V1 and V2 are the corresponding pathogen genes for avirulence, the interaction between host and pathogen is somewhat more complicated (Table 4.4). Of particular importance to remember is that resistant reactions conditioned by one gene-for-gene relationship are generally epistatic over susceptible reactions which might have resulted from other combinations of host and pathogen genes. Thus, in the interaction between pathogen genotype V1–v2v2 and host genotype R1–r2r2, the combination of v2v2 with r2r2 would give a susceptible reaction in a single gene interaction. However, the interaction between V1–

Table 4.4 *Genetic interactions in the gene-for-gene model involving two genes in both the host and a diploid pathogen. Values in bold type are the overall response of particular host/pathogen genotype combinations; values in parentheses give the particular response resulting from each pair of resistance and avirulence gene combinations, respectively.* (R *and* r, *alleles for resistance and susceptibility;* V *and* v, *alleles for avirulence and virulence;* −, *resistant reaction;* +, *susceptible reaction.*)

Pathogen genotypes	Host genotypes *R1–R2–*	*R1–r2r2*	*r1r1R2–*	*r1r1r2r2*
V1–V2–	− (−/−)	− (−/+)	− (+/−)	+ (+/+)
V1–v2v2	− (−/+)	− (−/+)	+ (+/+)	+ (+/+)
v1v1V2–	− (+/−)	+ (+/+)	− (+/−)	+ (+/+)
v1v1v2v2	+ (+/+)	+ (+/+)	+ (+/+)	+ (+/+)

and R1− gives a resistant reaction that is epistatic to this susceptible reaction. As a consequence, the interaction between pathogen and host genotypes V1–v2v2 and R1–r2r2 results in an overall resistant reaction in this two-gene model.

The pattern described above is the most common, but not the only, form that gene-for-gene interactions may take. With pathogens that produce host-specific toxins, the interaction is the reverse (Leonard, in press, *a*). In the case of the *Avena sativa–Helminthosporium victoriae* (seedling blight) host–pathogen system, a susceptible reaction is produced only when a pathogen genotype that produces the toxin is combined with a dominant gene for sensitivity in the host (Ellingboe, 1976). Similar gene-for-gene interaction patterns occur in diseases like southern corn leaf blight and leaf spot of maize caused by *Bipolaris maydis* and *B. zeicola*, respectively (Leonard, 1984).

Originally the gene-for-gene hypothesis was developed from the parallel application of genetic studies to both host and pathogen. Unfortunately, however, in many host–pathogen interactions the complete life cycle of the pathogen is not known (for example, *Puccinia striiformis* (stripe rust)) or is difficult to manipulate (for example, *P. graminis tritici* (wheat stem rust)). Person (1959) argued that gene-for-gene relationships should occur as a general rule and gave a comprehensive description of the features of such interactions. This allowed the development of ways of testing for gene-for-gene relationships even when genetic data were not available. Using a combination of the approaches devised by Flor and Person, gene-for-gene relationships are now *known* or *postulated* to occur in over 25 host plant–pathogen systems (Table 4.5).

Table 4.5 *Host–pathogen interactions in which gene-for-gene relationships have been demonstrated or suggested to occur (after Person & Sidhu, 1971; Day, 1974; Sidhu 1980)*

Host–pathogen system	Reference
Bunts	
Triticum–Tilletia caries	Metzger & Trione, 1962
Triticum–Tilletia contraversa	Holton, Hoffman & Duran, 1968
Mildews	
Hordeum–Erysiphe graminis hordei	Moseman, 1959
Triticum–Erysiphe graminis tritici	Powers & Sando, 1960
Lactuca–Bremia lactucae	Crute & Johnson, 1976
Rusts	
Avena–Puccinia graminis avenae	Martens, McKenzie & Green, 1970
Coffea–Hemileia vastatrix	Noronha-Wagner & Bettencourt, 1967
Glycine–Phakopsora pachyrhizi	Burdon & Speer, 1984
Helianthus–Puccinia helianthi	Sackston, 1962
Linum–Melampsora lini	Flor, 1942
Triticum–Puccinia graminis tritici	Luig & Watson, 1961
Triticum–Puccinia recondita	Samborski & Dyck, 1968
Triticum–Puccinia striiformis	Zadoks, 1961
Zea–Puccinia sorghi	Flanges & Dickson, 1961
Smuts	
Avena–Ustilago avenae	Holton & Halisky, 1960
Hordeum–Ustilago hordei	Sidhu & Person, 1972
Triticum–Ustilago tritici	Oort, 1963
Other fungi	
Cucumis–Fusarium oxysporum melonis	Risser, Banihashemi & Davis, 1976
Lycopersicon–Cladosporium fulvum	Day, 1956
Malus–Venturia inaequalis	Boone & Keitt, 1957
Oryza–Pyricularia oryzae	Kiyosawa, 1980
Phaseolus–Colletotrichum lindemuthianum	Albershiem, Jones & English, 1969
Solanum–Phytophthora infestans	Black *et al.*, 1953
Solanum–Synchrytrium endobioticum	Howard, 1968
Bacteria	
Gossypium–Xanthomonas malvacearum	Brinkerhoff, 1970
Viruses	
Lycopersicon–tobacco mosaic	Pelham, 1966
Lycopersicon–spotted wilt	Day, 1960
Solanum–potato virus X	Howard, 1968

To date, virtually all gene-for-gene systems have involved hosts that are economic crops. This has led several workers (Day, 1974; Sidhu, 1980; Day, Barrett & Wolfe, 1983; Barrett, 1984, and in press) to question the general relevance of the concept and suggest that it might be an artefact resulting from intensive breeding for disease resistance in agriculture. These authors argue that the standard design of breeding programmes typically favours resistance that is simply inherited, dominant in expression and that has a major phenotypic effect. On the other hand, resistance that is multigenic, that is recessive or whose effect is to reduce epidemic rates despite producing an apparently susceptible reaction, is likely to be discarded during early selection. As a consequence, the relative importance of gene-for-gene interactions in the entire gamut of disease resistance mechanisms in plants has become inflated. Furthermore, these authors have pointed out that for many of the proposed gene-for-gene systems listed in Table 4.5, genetic analyses have been restricted to studies of the host. As more becomes known about the genetic basis of avirulence, additional examples of departures from a strict one-to-one relationship between resistance and avirulence genes are likely to become apparent.

While these sentiments undoubtedly provide a timely reminder that gene-for-gene interactions are only one expression of genetically controlled disease resistance in plants, evidence is becoming available to suggest that the overwhelming representation of agricultural examples in Table 4.5 results in large part from the preoccupation of plant pathologists with agricultural systems. Circumstantial evidence supporting a wider occurrence of gene-for-gene systems is found in a number of wild plant–pathogen interactions where races of pathogens can be defined in terms of a range of race specific resistance genes occurring in individual host populations (see Chapter 7).

From the ecologist's point of view it is interesting to note that the species for which gene-for-gene relationships have been demonstrated or suggested to occur represent a wide range of host and pathogen genera. Amongst the hosts there is a very strong representation of annual grasses but examples of annual (*Linum* sp., *Helianthus* sp.) and perennial (*Solanum* sp.) herbs, shrubs (*Coffea* sp.) and trees (*Malus* sp.) also occur. Undoubtedly this list will expand as further studies are made.

Applications to the analysis of disease resistance in wild plant populations

Genetic analyses of the interactions occurring between host and pathogen populations in non-agricultural situations are hampered both by the diversity of resistance mechanisms and by complications introduced by the variable genetic background against which these are likely to be expressed. As a consequence, in the foreseeable future, studies of race non-specific resistance in such populations are likely to be restricted to the documentation of those aspects of resistance which can be monitored without recourse to elaborate techniques. The determination of inter- and intra-populational variation in factors like the latent period, infectious period and the number of productive lesions occurring per unit area are all fairly simply determined and have even been attempted on a few occasions (see Chapter 7). Alternatively, in some host–pathogen systems, variations in the combined expression of a complex of race non-specific resistance mechanisms may be measured as differences in levels of disease severity occurring during the development of pathogen epidemics. While the genetic basis of such differences in natural populations has yet to be determined, the difficulties do not seem insurmountable.

Detailed consideration of potentially important (but difficult to monitor) epidemiological factors such as minor variations in the size of sporulating lesions or the absolute number or rate of spore production are likely to remain intractable to accurate assessment on a population basis. Equally, the polygenic nature of many race non-specific resistance mechanisms makes it very unlikely that a genetic analysis of such resistance occurring within even a single wild plant population will be attempted for some time. However, race non-specific resistance mechanisms are undoubtedly an important part of the defensive armoury of plants occurring in non-agricultural systems. Although this type of active disease resistance is difficult to quantify, it should not be dismissed as unimportant.

In comparison to race non-specific resistance, race specific resistance is amenable to study in wild plant populations. The expression of such resistance in seedlings derived from a random sample of plants growing in wild populations has been documented on many occasions (see Chapter 7). Because of complications introduced by the simultaneous presence of race non-specific resistance and variations in the genetic background of the host, such analyses may show a continuum of infection type responses rather than a number of discrete classes. However, when the resistance of such populations is assessed with two or more different pathogen races, marked changes in infection types are often encountered. From a knowledge of the principles of the gene-for-gene hypothesis, the minimum number of different resistance genes responsible for these differential responses can then be estimated.

As a simple example of this we shall analyse the hypothetical data

Table 4.6 *Hypothetical data showing the response of a theoretical host plant population to six different races of a fungal pathogen* (–, *resistant reaction*; +, *susceptible reaction*)

Host genotype	Pathogen race P1	P2	P3	P4	P5	P6
H1	+	+	+	+	+	+
H2	+	+	+	+	–	+
H3	+	+	–	+	+	–
H4	+	+	+	–	–	–
H5	+	+	–	–	–	–
H6	+	–	–	–	–	–

presented in Table 4.6, showing the general category of resistance response of individual members of a theoretical host population to six different races of a fungal pathogen. First, take the host plant H1. In all cases the infection type resulting from the inoculation of this plant with each race of the pathogen was classified as susceptible. Hence no race specific resistance functional against these particular pathogen races was present.

The simplest explanation of the pattern of resistance and susceptibility shown by the second host genotype (H2) is the presence of a single resistance gene effective against pathogen race P5. This race is, therefore, avirulent to that resistance gene. Similarly, the response of the third host line (H3) can be explained by postulating the presence of another gene that confers resistance against races P3 and P6. Because the fourth host line (H4) shows a pattern of resistance and susceptibility which has some similarities with genotype H2 (both lines are resistant to pathogen race P5) but some differences (H4 is resistant to races P4 and P6 to which H2 is susceptible) we can propose two solutions. Either a single resistance gene that confers resistance to races P4, P5 and P6 is present in host genotype H4 or this line contains two resistance genes – one of which confers resistance to race P5 (the same gene as in H2) and one which is effective against races P4 and P6. Taking a conservative approach we would initially accept the former explanation. When a host like H5 responds to the six pathogen races with a pattern of resistance and susceptibility which may be obtained by a simple summation of the responses of other lines (H3 and H4) then a conservative approach would argue against the presence of yet another resistance gene without further genetic analysis. Finally, the resistant response of host H6 to all but race P1 of the pathogen, and in particular its response to P2, again argues for the presence of at least one extra resistance gene.

Applying these arguments, a minimum of four resistance genes (single genes in H2, H3, H4 and H6) is needed to explain the patterns of resistance and susceptibility observed in this hypothetical plant population. (Many other resistance genes *may* be present, however.) In a more complex real situation (Burdon, unpublished data) this approach predicted a minimum number of five resistance genes. Genetic analyses involving crossing between lines demonstrated that six different genes were actually present. All but one line carried a single resistance gene; the exception carried two.

A similar argument can be used to estimate the number of avirulence genes present in the pathogen population. In this analysis we can ignore host line H1. The susceptibility of this line to all six races of the pathogen can result either from a complete correspondence of virulence in the pathogen with resistance in the host or from a total lack of resistance genes in this host. As these two possibilities cannot be distinguished phenotypically, a prudent assessment argues for adoption of the latter, simpler explanation. Now, considering individual pathogen races we find that the simplest explanation of the pattern of resistance and susceptibility induced by race P2 is the presence of a single avirulence gene which conditions the incompatible (resistant) reaction on host line H6. In a similar way the unique patterns of susceptibility and resistance induced by races P3, P4 and P5 can most conservatively be interpreted to indicate the presence of three further distinct avirulence genes. Finally, the pattern produced as a result of the interaction of race P6 and host lines H1 to H6 do not require the postulated presence of any additional avirulence genes. Rather a combination of the avirulence of races P3 and P4 would produce an identical pattern. Alternatively, the patterns obtained *could* reflect various combinations of these and yet *other* avirulence genes. However, we do not require this level of complexity to explain the results!

The various approaches outlined here are beginning to be applied in wild plant and pathogen populations. Ultimately this will allow the extent and role of various disease resistance mechanisms in such systems to be assessed while, at the same time, pathogen variability is documented. This area of the interaction between host and pathogen is considered in Chapter 7.

5

Environmental modification of host–pathogen interactions

The occurrence and expression of disease in plants is not determined solely by the genetic constitutions of the host and pathogen involved. Like all other biological systems, the phenotypic expression of resistance or susceptibility in the host and avirulence or virulence in the pathogen may be modified markedly by variations in the abiotic and biotic environment. This dependence of the level of disease and the extent of symptom expression on the interaction of host, pathogen and environment has long been recognized in pathology in the concept of the disease triangle (Figure 5.1).

There are three basic ways in which the environment may affect any interaction between a pathogen and its host: (1) by acting directly on the pathogen itself during those periods when it is not intimately linked to its host, changes in the environment may reduce survival or even prevent the establishment of new parasitic relationships; (2) environmental variation may induce substantial changes in the relative susceptibility or 'predisposition' (Yarwood, 1959) of a host plant to infection; and (3) once a parasitic relationship is established, variations in environmental conditions may result in variations in the expression of disease symptoms. Behind each of these basic interactions are a myriad of ways in which environmental variables may influence the occurrence or expression of disease. On a broad spatial scale, patterns of disease occurrence are most likely to be correlated with climatic features which influence the survival of pathogens. As the spatial scale reduces to a regional and then a local level, however, the relative importance of the many factors that may affect host–pathogen interactions alters. The emphasis changes from a simple effect of the environment on the pathogen alone towards a complex array of effects on the host and the host–pathogen association itself.

It is not the intention of this chapter to provide an exhaustive or comprehensive treatment of these mechanisms – such a task would be

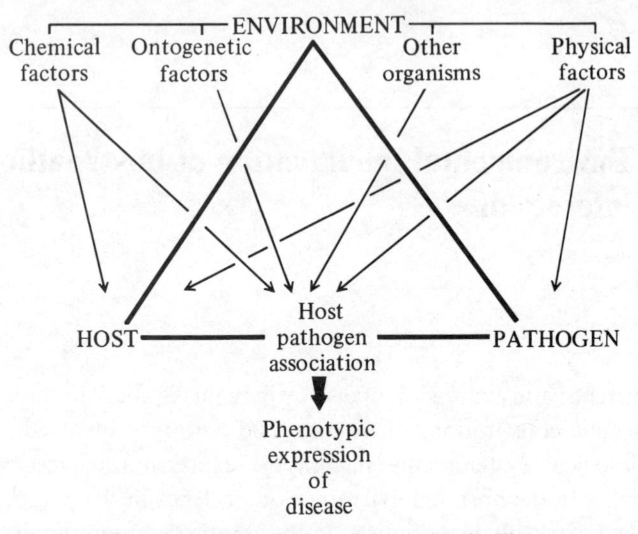

Figure 5.1 The basic concept of the disease triangle. The occurrence and expression of disease results from an interaction of host, pathogen and environment. The main features of the environment and their primary site of action are superimposed on the basic model.

monumental and would result in a chapter longer than the present volume! Rather, this chapter aims to stress the major importance of the environment as a modifying influence on host–pathogen interactions in the field. Not only are environmental effects important in determining the geographical distribution of disease but even within one habitat subtle fluctuations in the environment may be reflected in variations in the extent and nature of disease symptoms. This variation is particularly relevant to considerations of the relative severity of disease in agricultural and non-agricultural systems. Equally, such environmental modification of symptom expression has important implications for the analysis of population structure – phenotypic differences observed between members of a plant population in the field cannot be attributed to differences in genotype without further analysis.

The effect of environment on the spatial distribution of disease

Macro-geographical variations in disease occurrence

While some pathogens appear to be capable of causing disease over the entire distribution range of their hosts, the majority are somewhat less tolerant of environmental variation and are restricted to some degree or other to specific parts of their host's range (Weltzien, 1972; Colhoun,

1973). The two most important environmental factors influencing the overall geographical distribution of individual pathogens are temperature and moisture. Frequently the final pattern of disease occurrence observed in the field results from an interaction of these two factors.

These variables affect the occurrence of disease primarily through their direct effect on the survival of inoculum during its passage from one host to another and on the occurrence of conditions suitable for subsequent spore germination. In some instances the restrictions these climatic factors place on the distribution of a pathogen are quite marked. Thus, in North America, cotton anthracnose (*Glomerella gossypii*) occurs only spasmodically in regions receiving less than 25 cm of rain in summer but regularly in those receiving more than 30 cm (Miller, 1966). The occurrence of many other plant pathogens, for example *Pseudomonas citri* (bacterial canker of citrus) (Reichert & Palti, 1967) and *Spongospora subterranea* (potato scab) (Weltzien, 1972) is also particularly favoured by high levels of precipitation. In some of these cases high rainfall *per se* may favour the spread of disease (for example, in a splash-borne pathogen). In many other cases, however, increasing rainfall results in higher humidity and changes in soil water status. These conditions often favour infection. Conversely, some diseases appear to be favoured by low rainfall. Atkinson & Grant (1967) found that the amount of damage caused by wheat streak mosaic virus to the 1964 Albertan (Canada) wheat crop was strongly correlated with precipitation levels. Most damage was restricted to areas which received less than 5 cm of rain the previous spring. It seems likely that the ultimate distribution of disease in this case resulted from a direct effect of the environment on the survival of the virus vector.

Similar broad-scale differences in disease occurrence have been correlated with temperature. *Urocystis cepulae* (onion smut) is regularly distributed throughout the United States on infected onion sets. However, the disease is only found in northern areas where soil temperatures remain low enough to favour pathogen development throughout the growing season (Walker & Wellman, 1926; Colhoun, 1973). Similarly, soil temperatures above 23 °C limit the spread of *Synchytrium endobioticum* (potato wart) and thus reduce the occurrence of this pathogen in some parts of the world (Reichert & Palti, 1967).

Micro-geographical variations in disease occurrence

Even within a region that is generally favourable for the growth and reproduction of specific pathogens, marked changes in disease occurrence and severity may occur as a result of micro-environmental changes over very short distances. At this scale, the level of disease occurring in the field is likely to be determined not only by the direct effect of the environment on the pathogen but also by features associated with the direct effect of the environment on the host (for example, differences in soil nutrition) and on the parasitic association itself. This inevitably leads to a fine mosaic of environmental conditions which find expression in variations in disease occurrence or severity. Here we are particularly concerned with those features of the topography of a region that may affect the occurrence of conditions suitable for the germination of pathogen spores.

Most meteorological parameters are affected by local topographical features like hills and valleys. Variations in aspect, slope and degree of exposure interact with temperature, moisture, wind and insolation levels to produce highly localized differences in micro-climatic conditions (Geiger, 1965). Relative humidities tend to be higher in valley bottoms than on neighbouring exposed high ground and, consequently, dew periods tend to be shorter in the latter situation. Similarly, easterly facing slopes tend to have shorter dew periods than do westerly or southerly facing slopes (northerly facing in the northern hemisphere!). Incident solar radiation levels are markedly lower on polar-facing than on equatorial-facing slopes. The potential effects of these micro-climatic differences on the germination, establishment and later occurrence of disease has been demonstrated on many occasions in agricultural situations (Palti, 1981). In a somewhat less regimented situation, differences in light intensity have been implicated as a causal factor in the spatial distribution of a die-back disease of Corsican pine in north-east England (Read, 1968). Die-back caused by the fungus *Brunchorstia pinea* was most common on north-facing slopes where low winter light levels predisposed host plants to infection.

The effect of differences in slope and aspect on the botanical composition of adjacent sites has long been recognized. It is perhaps not surprising, therefore, that similar differences should occur in disease occurrence. However, even more subtle variations in disease levels have been shown to result from the effects of shading. Palti & Netzer (1963) have described a study of a tomato crop attacked by *Phytophthora infestans* (blight) in which the shadow cast by a row of trees bordering the southern side of the crop was reflected in higher levels of disease in the shaded areas than elsewhere in the field. A similar phenomenon was noticed during the development of an epidemic of southern corn leaf blight (caused by *Helminthosporium maydis*) in maize fields in Connecticut, United States

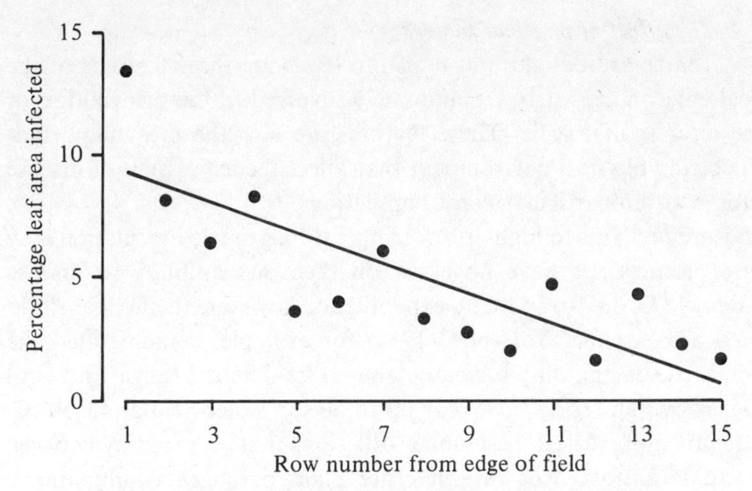

Figure 5.2 The incidence of southern corn leaf blight (*Helminthosporium maydis*) in relation to the relative distance of plants from trees growing along the south-eastern edge of a field. Row 15 is in open sun in the middle of the field. The inverse linear relation is described by the equation $y = 9.8 - 0.6x$. (Redrawn from Lukens & Mullany, 1972.)

(Lukens & Mullany, 1972). Within affected fields the disease was more severe in areas shaded by trees than those in open sun. Moreover, rows of maize closest to the trees, and hence those shaded for the longest periods of time, were the most heavily diseased (Figure 5.2).

The effect of environment on disease occurrence and expression within individual host stands

The environment affects the expression of disease symptoms mainly through its effect on the host prior to infection (predisposition) and on the host–pathogen association once infection has occurred. While such interactions may be responsible, in part, for differences in the severity of disease between adjacent host stands, their greatest relative effect is found within individual plant communities or populations. There, the fine-scale disposition of one plant relative to another may make a considerable difference in the degree to which an individual is predisposed to attack by a pathogen. The size of individuals and their neighbours, differences in the degree of shading of individual leaves, in the mineral nutrition of plants and in the ontogenetic age of different plant parts, and the presence or absence of other pathogenic and non-pathogenic organisms may all affect the occurrence and expression of disease.

The effect of physical factors

Temperature, light and moisture levels are three features of the physical environment that are intimately involved in the distribution of disease on a spatial scale. These features are also the most important aspects of the physical environment that affect the expression of disease symptoms within particular plant populations.

Exposure of plants to high or low temperature prior to inoculation may increase, decrease or have no effect on their susceptibility to disease (Colhoun, 1973, 1979). In many experiments, however, the temperatures involved are extreme. Yarwood (1956) for example, demonstrated the heat-induced susceptibility of bean plants to six different fungal and viral pathogens by immersing leaves for up to 60 s in water held at 45–50 °C. This treatment increased the number of lesions that occurred by between two- and seven-fold. For several other plant–pathogen combinations, though, no noticeable effects were detected. The practical significance of results obtained with such temperature treatments is obscure.

More modest and ecologically plausible temperatures are also able to affect disease susceptibility in some plants. Rice blast lesions caused by the pathogen *Pyricularia oryzae* are considerably larger on seedlings grown at low night temperatures (20 °C) than on those grown at higher ones (30 °C) (Ramakrishnan, 1966). Conversely, Kassanis (1952) found a marked increase in the susceptibility of *Nicotiana glutinosa* to five different viruses if plants held for as little as 6 h at 36 °C were promptly returned to a control temperature of approximately 20 °C following inoculation. However, this increased susceptibility did not necessarily occur if inoculated plants remained at the higher temperature. In fact, in the case of infections of tobacco necrosis, tomato bushy stunt and cucumber mosaic virus, no disease lesions were produced following this treatment. For infections of tomato spotted wilt and tobacco mosaic virus, 10–90% fewer lesions occurred than on control plants. More interestingly, the symptoms shown by plants infected with tobacco mosaic virus at 36 °C were quite different to those of plants infected at 20 °C. At 36 °C local lesions were chlorotic rather than necrotic in appearance and the plants became systemically infected although this was not visually apparent. When such individuals were moved to the lower temperature the entire plant rapidly died. This temperature-related change in the visual appearance of plants that are systemically infected with viruses has been reported on a number of occasions (Gill & Westdal, 1966).

Once an infection has become established, temperature variations may also affect the speed of development of the host–pathogen association directly. This then becomes apparent as changes in the length of the latent and incubation periods. Observations of this nature are particularly common in studies involving foliar diseases like rusts and mildews. These diseases also provide examples of marked temperature-related changes in the disease symptoms conditioned by specific resistance genes. The most extreme cases of such variation in symptom expression are illustrated by a number of genes for stem and leaf rust resistance in wheat (e.g. *Sr6*, *Sr15* and *Lr20*). These resistance genes are thermolabile. At temperatures above 26 °C (*Sr6*, *Sr15*) and 30 °C (*Lr20*), they are inoperative. At lower temperatures these genes are increasingly effective against avirulent races of the appropriate pathogen. In the case of the *Sr15* gene, plants held at 15 °C after inoculation with avirulent races of *Puccinia graminis tritici* (stem rust) produce a typical resistant reaction characterized by a mixture of hypersensitive flecks and very small uredia surrounded by necrotic tissue (Figure 5.3). Plants carrying the *Sr15* gene and held at 20 °C show a slightly enhanced infection type. In contrast, plants held at 26 °C following infection show a fully susceptible response characterized by very large pustules with no chlorotic zones (Figure 5.3; R. A. McIntosh, personal communication).

Changes in light intensity and moisture regimes may similarly affect the expression of disease symptoms. This may occur through their direct action on hosts and pathogens or indirectly through interactions with each other or with temperature fluctuations. Shading or dark treatments of plants prior to inoculation increases the susceptibility of a range of species to a variety of fungal and viral pathogens (Bawden & Roberts, 1947; Yarwood, 1959; Read, 1968). Variations in light intensity also affect the relative success of spore germination and the establishment of infections. Urediospores of various rust fungi react quite differently to light during germination. Those of *Puccinia coronata* (crown rust) and *P. graminis avenae* (stem rust) germinate most rapidly in the dark (Kochman & Brown, 1976*a*) while those of *Uromyces appendiculatus* (bean rust) are apparently unaffected (Snow, 1964). Subsequent host penetration is unaffected by increasing light intensities in *P. coronata* but is enhanced in *P. graminis avenae* (Kochman & Brown, 1976*b*).

Once a parasitic association is established, light intensities similar to those that might occur as a result of the shading of one leaf by others may affect the expression of disease symptoms. Pandey & Wilcoxson (1970) found that the symptoms produced by the pathogen *Leptosphaerulina brissiana* (leaf spot) on lucerne were of a resistant type at low light levels but became more susceptible as light intensity increased. Such changes are

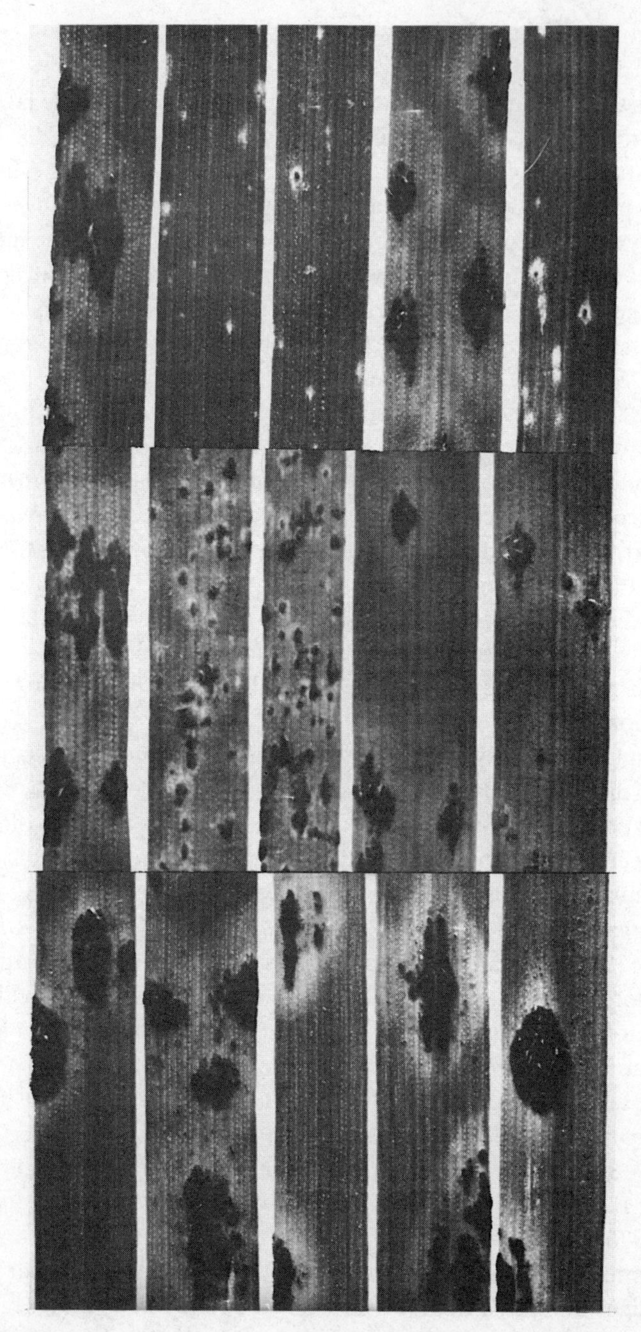

Figure 5.3 Influence of temperature on infection types produced by lines with *Sr15* and *sr15* when infected with an avirulent race of *Puccinia graminis tritici*. From left to right, host lines are: W2691 (*sr15*), Thew (*Sr15*), Norka (*Sr15*), Chinese Spring (*sr15*), Chinese Spring*6/Axminster 7A (*Sr15*). Temperatures running from top to bottom are: 15 °C, 20 °C, 26 °C. (Photographic plate courtesy of Dr H. D. M. Gousseau and Mr D. J. S. Gow, University of Sydney, Plant Breeding Institute, Castle Hill.)

typically found in the symptoms expressed on graminaceous hosts infected with rust pathogens.

For most bacterial and fungal diseases, moisture levels are critically important during the vulnerable stages of spore germination and penetration, but are of less importance either prior to infection or during the subsequent development of the parasitic association. This is not to imply, however, that either drought or flooding does not affect the predisposition of plants to infection by various pathogens. Periods of drought have been implicated in the subsequent occurrence of many decline, blight and die-back diseases of trees, although the precise means by which this occurs is unclear (Schoeneweiss, 1975). In fact, diseased plants in general appear to be less able to withstand unfavourable environmental conditions, whether these are expressed as extremes of temperature or water availability, than do healthy ones.

The effect of nutritional factors

Plant nutrients may affect the occurrence and expression of disease in a number of ways. Through their effect on plant vigour, nutrients may alter the micro-climate of a developing plant stand or, in the case of seedlings, determine whether individual hosts can outgrow potential pathogens (see also 'the race to maturity', Chapter 2). As a result of alterations to the soil environment, nutrients may affect the survival and development of root pathogens. Finally, through their effect on the basic biochemistry of host plants, nutrients may affect the development of disease symptoms (Colhoun, 1973).

Of the major plant nutrients, most is known about the effects of nitrogen, phosphorus and potassium. Variations in nitrogen availability markedly affect disease levels, although the form in which nitrogen is supplied – ammonium or nitrate nitrogen – is of considerable importance. The effects of either of these forms of nitrogen depends on many factors and are not the same for all host–pathogen associations. In a review of over 35 specific host–pathogen associations, Huber & Watson (1974) found that ammonium and nitrate nitrogen consistently had opposite effects on disease levels. If applications of nitrate nitrogen reduced disease, then applications of ammonium nitrogen increased it and *vice versa*. For example, nitrate nitrogen decreases the severity of *Fusarium* wilt of tomatoes; ammonium nitrogen increases it. Conversely, the severity of *Verticillium* wilt is reduced by the ammonium form of nitrogen but increased by the nitrate form.

Plants deficient in either phosphorus or potassium tend to be highly susceptible to many root and foliar pathogens (Graham, 1983). As the availability of these nutrients increases, the level of disease declines. Supra-optimal concentrations, however, appear to have little effect. At such levels, increases, decreases and no change in disease levels have all been reported (Jenkyn & Bainbridge, 1978).

Deficiencies of most minor nutrients generally result in plants that are more susceptible to a variety of diseases. Some deficiencies result in greater susceptibility to a wide range of pathogens; the effects of others appear to be more restricted. Deficiencies of manganese and nickel are particularly associated with greater susceptibility to powdery mildews and rusts, respectively (Graham, 1983). Deficiencies of boron and copper, on the other hand, induce increased susceptibility to a wide range of foliar and root pathogens. Applications of boron increase resistance to powdery mildew of sunflowers (Butler & Jones, 1955), *Fusarium* wilt of flax (Keane & Sackston, 1970) and *Synchytrium endobioticum* (wart) of potatoes (Hampson, 1980). Similar additions of copper reduce the occurrence of powdery mildew of barley (*Erysiphe graminis hordei*), ergot (*Claviceps purpurea*) on rye and *Gaeumannomyces graminis* (take-all) on wheat (Graham, 1983). These observations contrast with the ambiguous results obtained when zinc is added to deficient plants; in some instances beneficial effects have been reported but in others either no effect was observed or increases in susceptibility have occurred (Graham, 1983).

Overall, most major and minor plant nutrients have at least some effect on the susceptibility of hosts to plant pathogens, although the literature is peppered with results of contradictory nature. These inconsistencies often reflect interactions between the effects induced by different nutrients.

The effect of ontogenetic factors

The response of individual plants to particular pathogens often changes through the life of the plant. Thus plants are more susceptible to damping-off pathogens during the germination and early establishment phase of growth than they are as adults (see Chapter 2). Equally, the susceptibility of specific organs may change as they age. Leaves of *Populus nigra*, on the other hand, show an age-related sequence of increasing and then decreasing susceptibility to rust caused by *Melampsora larici-populina* (Sharma, Heather & Winer, 1980). On individual shoots, leaf maturity and consequently leaf position on the shoot is a significant factor in the susceptibility of leaves to attack. Once shoots are more than 8 days old a basipetal sequence of zones of susceptibility is established (Figure

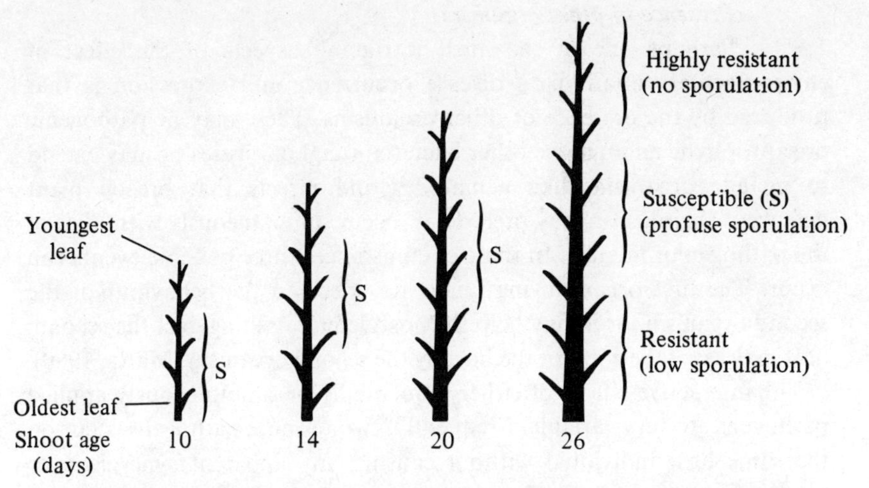

Figure 5.4 Ontogenetic changes in the susceptibility of successively formed leaves of *Populus nigra* to infection by *Melampsora larici-populina* (leaf rust). (Redrawn from Sharma *et al.*, 1980.)

5.4). The youngest leaves are virtually immune, intermediate-aged leaves are highly susceptible and, finally, the oldest leaves are again resistant.

Overall, in addition to situations where the level of resistance or susceptibility remains unaltered with age, four categories of ontogenetic response to pathogens have been recognized, *viz*: increasing resistance with age; increasing susceptibility with age; relatively greater resistance during the mid-life period; and, finally, relatively greater susceptibility during the same period (Yarwood, 1959).

Although ontogenetic changes in the relative susceptibility or resistance of plants to pathogen attack are common, few reliable generalizations can be made (Yarwood, 1959). However, it does seem likely that such differences may play an important role in reducing the potential of natural plant stands to maintain the epidemic development of a pathogen. Unlike most agricultural crops that are even-aged, wild plant populations are frequently very variable in their age structure. At any given time the actual density of susceptible host tissue in such stands is often considerably lower than the apparent density. In agricultural populations, on the other hand, the actual and apparent densities are likely to be the same.

Presence of other organisms

Perhaps one of the most intriguing aspects of the effect of environmental variation on disease occurrence and expression is that produced by the presence of other organisms. These may be pathogenic or saprophytic and may be other bacteria, fungi or viruses or may extend to include organisms like nematodes and insects that breach plant defences. Their action may precede, or occur simultaneously with, that of the pathogen of interest. In these circumstances three possible events can occur. The first organism may have no effect on the behaviour of the second organism, or it may help the host defend itself against the second, or it may assist the entry of the host by the second organism (Matta, 1980).

The interactive effects of either sequentially or simultaneously applied pathogens are very variable. Frequently, two or more pathogens occur on the same host individual without causing any apparent synergistic or antagonistic interaction. This may occur regardless of the phylogenetic relationship between the organisms: when they are totally unrelated, when they are members of the same general group or when they are highly specialized races or forms of the one species. Equally, however, many of these combinations result in either enhanced or reduced pathogen attack and symptom expression.

Synergistic interactions have been reported on many occasions. Potato plants infected with potato virus X show increased susceptibility to *Alternaria solani* (early blight) (Hooker & Fronek, 1961); those infected with leaf roll virus show increased susceptibility to *Phytophthora infestans* (de Cubillos & Thurston, 1975). Wheat infected with *Erysiphe graminis* (powdery mildew) or *Pseudocercosporella herpotrichoides* (eyespot) is predisposed to attack by *Septoria tritici* (speckled leaf blotch) (Brokenshire, 1974) or *S. nodorum* (glume blotch) (Jones & Jenkins, 1978) and prior infection by barley yellow dwarf virus initially suppressed but subsequently enhanced the severity of *E. graminis* on oats and barley (Potter, 1980).

Antagonistic reactions in which resistance is induced in the host as a result of the presence of a pathogen have also been found extensively amongst fungal and viral pathogens. This phenomenon is also referred to as 'acquired immunity' or 'cross-protection' and in virus–virus combinations usually only occurs between related strains of the same virus (Gibbs & Harrison, 1976). In interactions between viral and fungal pathogens, barley, oat and wheat plants previously infected with barley yellow dwarf virus were more resistant to their respective rusts (*Puccinia hordei*, *P. coronata* and *P. recondita*) than were healthy plants (Potter, 1982). This induced resistance was expressed as a reduction in the number of pustules that developed. A similar reaction occurred when *P. coronata* was inoculated onto perennial ryegrass plants infected with ryegrass mosaic virus. Interestingly, barley yellow dwarf virus had no effect in this situation (Latch & Potter, 1977).

In fungus–fungus interactions most interest has centred on resistance induced by avirulent isolates of a pathogen against the subsequent attack of virulent isolates. This has frequently been shown in a range of rusts and powdery mildews (Johnson & Allen, 1975; Chin & Wolfe, 1984*a*). In general the resistance induced in this manner is localized to within a few millimetres of the reaction produced by the inducing pathogen (Johnson, 1978). (A marked exception to this has been found in the systemic resistance that is induced in cucumber plants subjected to a spatially limited infection of *Colletotrichum lagenarium* (anthracnose) (Kuc, Shockley & Kearney, 1975).) The ability of avirulent races of a pathogen to induce resistance against virulent ones is not just a laboratory oddity. Recently, Chin & Wolfe (1984*a*) demonstrated that such resistance may noticeably affect the development of epidemics of *Erysiphe graminis* on barley in the field (see also Chapter 3).

Saprophytic micro-organisms in the phylloplane and phyllosphere have also been shown to impede, on occasions, the action of pathogens. However, rather than affecting symptom expression, these organisms tend to reduce disease occurrence. The action of saprophytic soil micro-organisms is the basic mechanism behind the functioning of 'suppressive soils' that can significantly reduce the occurrence of soil-borne pathogens such as *Phytophthora* (Cook & Baker, 1983).

Concluding remarks

The modifying influence of the environment on any basic host–pathogen interaction is a real and potent phenomenon. Any consideration of the role that pathogens play in plant communities that does not, at least implicitly, recognize this fact is seriously flawed. Changes in a wide variety of geographical, climatic, nutritional, developmental or biotic aspects of the environment, either singly or, more realistically, acting in concert with one another, inevitably have a profound effect on the occurrence and phenotypic expression of disease (Figure 5.1). Even subtle changes in one or two aspects of the environment may eventually be translated into marked changes in disease severity or symptom expression. These variations have significant implications for the analysis of the genetic structure of wild plant populations and undoubtedly contribute to the actual levels of disease observed in wild plant populations.

6

The effect of pathogens on the size of plant populations

Most plant populations have the potential to increase in size until restrained by physical limitations of the environment. Where pathogens are present, however, the high-density, single-species stands which result from such unrestricted growth are particularly vulnerable to attack. If the effects of pathogens are sufficient to cause reductions in the size and density of host populations, the resultant, partly unoccupied, environmental resource niche may be colonized by other (more resistant) plant species or by novel resistant genotypes within the same host population. These two possibilities are not mutually exclusive. Rather, while invasion by other species is an almost inevitable consequence of the reductions in longevity, fecundity and competitive vigour which diseases may cause, this disease pressure also acts as a potent selective agent inducing changes in the genetic composition of host plant populations.

For simplicity, an artificial division of the effects of pathogens on their hosts into numerical and genetic components has been adopted here. The present chapter investigates the numerical changes that occur and the chapter following deals with genetic interactions. However, as we shall see, interactions at the intra- and inter-specific levels have the same essential features and it must always be borne in mind that genetic changes in particular occur against a background of fluctuations in the numerical size of individual populations.

Theoretical models

Consumer–host interactions

The majority of consumer–host organism interactions can be divided into two main classes: those in which the hosts are animals and those in which this role is taken by plants. As May (1982) observed, animal host–parasite (both macro- and micro-parasites), host–parasitoid and predator–prey interactions form a continuum of consumer–animal host organism interactions. Despite many differences, models of these interactions are characterized by a concern for the predatory or parasitic interaction alone. The effect of competition between hosts and between hosts and non-hosts on the outcome of the primary interaction is not considered. Standing apart from these, and distinguished by the marked importance of persistent and intense competitive effects, are plant–parasite interactions.

In contrast to the limited theoretical assessments available concerning plant–parasite interactions at the population level, analyses of consumer–animal host interactions have a long and distinguished history. The high level of sophistication of many of the latter models encourages a brief consideration of some of the conclusions arising from those with the greatest similarity to plant–pathogen interactions – models of animal host–parasite interactions. Classical predator–prey or insect host–parasitoid models are not considered here, as the fundamental differences between these and plant–pathogen interactions preclude all but the most superficial of comparisons.

The most relevant animal host–parasite models currently available are those developed by Anderson and May (Anderson & May, 1978, 1979, 1981; May & Anderson, 1978, 1979; Anderson, 1979). These models are fundamentally based on a simple mathematical description, originally formulated by Kermack & McKendrick (1927), of the spread of parasite epidemics through host populations. Unlike previous analyses, however, a wider view of the effect of parasites on host populations has been provided by extending this basic model to include the host population itself as a dynamic variable, with its own birth and death rates. Furthermore, Anderson and May have recognized and treated separately two groups of parasitic organisms (micro- and macro-parasites). While distinctly different mathematical models have been developed to accommodate differences in the relationship between these parasites and their hosts (for example, micro-parasites such as viruses, bacteria and some protozoa often induce lasting immunity in their hosts while macro-parasites like helminths and arthropods do not), both groups of models predict that such parasites are capable of regulating the size of host populations.

More specifically, if we look at the case of persistent infections induced by macro-parasites as being the most analogous to plant–pathogen interactions then stable co-existence of host–parasite associations is generally found to result from an interplay of opposing stabilizing and destabilizing forces (Anderson, 1978, 1979). At least four factors tend to favour the occurrence of stable associations and variously affect the equilibrium size of host populations, *viz.* over-dispersion of parasite numbers per host, density-dependent parasite mortality or reproduction, parasite-induced host mortality that increases faster than does parasite burden, and parasite-induced mortality rates that are more severe at high host densities. Conversely, other features, including parasite-induced reductions in host reproductive potential, may have a destabilizing effect because they restrict the range of conditions under which stability may occur.

Plant–pathogen interactions

In contrast to the impressive models of host–parasite population dynamics developed for animal systems, little interest has been shown in investigating the behaviour of similar plant–pathogen interactions. Traditionally, epidemiological studies of plant pathogens have concentrated on the dynamics of pathogen populations in relation to changes in environmental factors. Alternatively, they have tried to relate the effects of disease to final crop yield. Because the majority of agricultural species are annual plants grown in stands of uniform density that are reconstituted annually, the host population has been regarded as fixed. Moreover, the genetic uniformity of such populations has also precluded much interest in the effects of differential competition between adjacent infected and uninfected individuals on the short- and long-term effects of pathogens. In fact, the only models which attempt to address at least some aspects of these questions are found in the ecological literature.

In the simplest of these models, Janzen (1970) focussed attention on the effects of parasites (herbivorous insects and fungal pathogens) on the number of distribution of offspring around parent plants. Assuming that the number of propagules found around a parent would decline asymptotically with distance while, simultaneously, the possibility that any given propagule would escape attack increased to an asymptote, Janzen argued that recruitment of individuals into the host population would be concentrated at that distance from the parent where the product of these two factors was greatest. Ultimately however, the effect that parasites have on host populations is linked to their effect on the health of individual hosts. In Janzen's model, distorted spatial patterns of recruitment result from the assumption that parasites kill young seedlings. On the other hand, pathogens that only reduce plant vigour may affect the growth rate of all members of a population equally, while having little effect on its total size or spatial distribution. Unfortunately, although models which consider the effect of pathogens on individual host plant populations may be conceptually similar to the highly developed animal host–parasite models mentioned earlier, they fail to account for the potentially confounding effects of inter-plant competition.

Competition between adjacent plants may greatly exacerbate the detrimental effects of disease. Plants affected by disease are generally less able to compete effectively for environmental resources than are healthy ones. As a result, models which examine changes in the size of populations without addressing accompanying changes in co-occurring species are likely to underestimate the ultimate effects of pathogens.

Plant competition studies have been greatly influenced by the de Wit (1960) model which is based on the assumption that the yield of each component of a mixture is strictly proportional to its share of environmental resources. While this model enjoys reasonably wide applicability in interactions between different species or genotypes with constant relative competitive abilities (Trenbath, 1974), in situations where competitive abilities are frequency-dependent it is unable to predict either the short- or the long-term consequences of competition. However, a strong intuitive argument can be made to support the suggestion that host-specific pathogens may have significant frequency-dependent effects on their hosts and hence on the relative competitive abilities of the components of a mixture (Burdon & Shattock, 1980; Burdon, 1982). In essence this argument is based on the superimposition of a host–pathogen interaction on an existing competitive interaction between host and non-host plant species. In the absence of disease the host species has a superior competitive ability, while in the presence of the pathogen, interactions between the frequency of host plants, the severity of disease and the competitive pressure exerted by non-host individuals may result in the long-term stability of previously unstable mixtures.

Gates *et al.* (1986) have tested the validity of this hypothesis by developing a deterministic model based on three differential equations that describe how the amount of non-host, healthy host and infected host plant material varies with time. Although exact solutions for these equations cannot be obtained, they can be solved iteratively with respect to time for any choice of parameters. Not surprisingly, different combinations of values may result in markedly different solutions even over the space of a single generation of plants (Figure 6.1).

More interestingly, when the interactions between diseased and healthy plants illustrated in Figure 6.1 are projected through a number of consecutive generations, they are also found to differ in respect to their

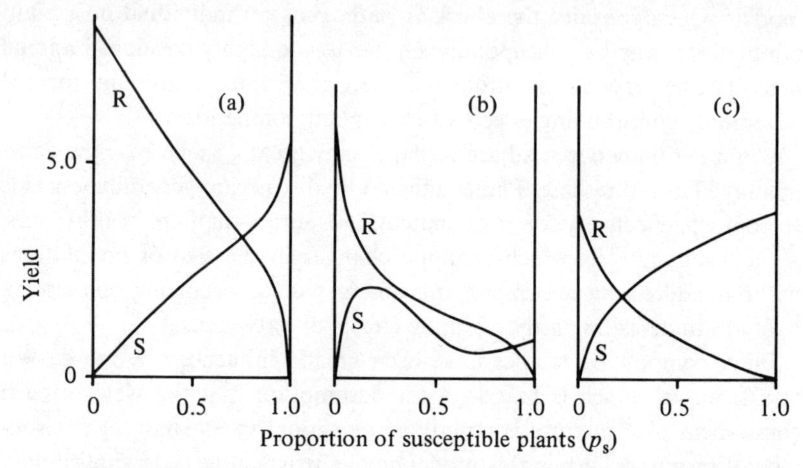

Figure 6.1 The effects within a single generation that a pathogen may have on the competitive interactions occurring between resistant (R) And susceptible (S) plants. (Redrawn from Gates *et al.*, 1986.) See text for explanations of (a), (b) and (c).

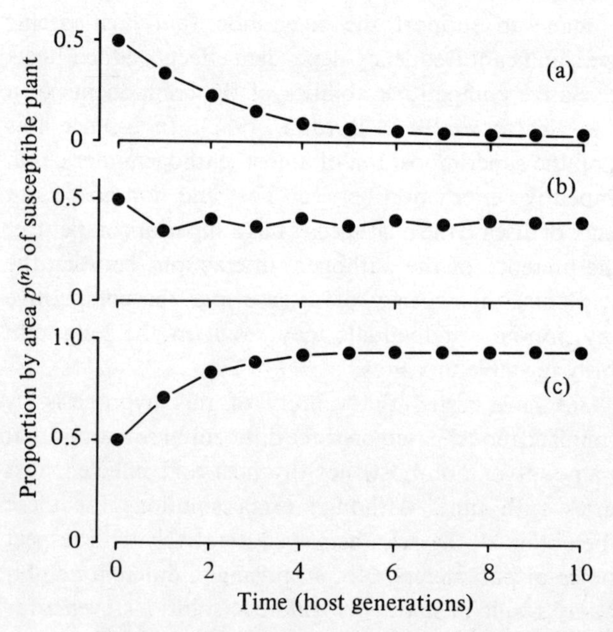

Figure 6.2 The effects over many generations of the competitive interactions occurring between resistant and susceptible plants. (Redrawn from Gates *et al.*, 1986.) See text for explanations of (a), (b) and (c).

long-term behaviours (Figure 6.2). Cases (a) and (c) are predictably unstable with the non-host and host species, respectively, excluding the other component from the mixture. Case (a) deals with the effect of a debilitating pathogen that severely reduces the competitive ability of a host species that was only marginally superior to the co-occurring non-host species. Inevitably the non-host species will eventually completely dominate this interaction. Similarly, the imposition of the effects of an ineffective pathogen on the interaction occurring between a highly competitive host and a weakly competitive non-host species (case c) is unlikely to be sufficient to redress the uneven competitive balance. The mixture is unstable and the host species triumphs. On the other hand, the intermediate case (b) is representative of a variety of long-term associations between host and non-host plant species that are made possible by marked frequency-dependent interactions between the host and the pathogen. The path followed during the establishment of the particular association illustrated is an oscillatory convergence to an equilibrium mixture. However, monotonic convergence to an equilibrium mixture, oscillatory convergence to a stable cycle or an essentially random or chaotic fluctuation in host frequency from generation to generation are all possible long-term solutions. The particular trajectory followed depends upon a range of parameters affecting epidemiology of the pathogen and the performance of the plant species involved.

The basic biological explanation for the frequency-dependent behaviour found in case (b) lies in a complex interaction between the size of the pathogen population and the relative density of host and non-host individuals (see Chapter 3 for a full discussion). When the relative frequency of host plants in the mixture is high, conditions for the rapid increase and spread of the pathogen are maximized, and the effect of disease on the competitive ability of affected hosts is marked. The competitive ability of host plants falls below that of the associated non-host plants. On the other hand, when host plants are present at low frequency, individual hosts are well spaced and interspersed with non-host plants. This leads to a drop in the size of the pathogen population and the effect of the resultant disease is much less than that occurring at high host frequencies. The competitive advantage of the host over the non-host species, while being reduced, still remains positive. Taken over all frequencies this type of interaction should produce a frequency-dependent effect.

A greater degree of realism may be injected into models of this type if one accepts that different sets of host-specific pathogens are likely to be present on both components of a mixture. In a highly simplified continuous systems model of such a scenario, Chilvers & Brittain (1972) showed that stable equilibrium mixtures might be achieved as a result of the depredations of such pathogens. Unfortunately, however, these authors did not attempt to analyse the conditions under which such stability might be achieved. It does seem likely, though, that the combined action of both sets of host-specific pathogens would strengthen frequency-dependent interactions.

Stability is clearly not an inevitable consequence of the differential interaction of pathogens with the plant genotypes occurring within an area. Rather, the long-term outcome of interactions between diseased and healthy plants depends greatly upon the growth rates and relative competitive abilities of the plants involved, on the effect of the pathogen on its hosts and on its transmission efficiency (Gates *et al*., 1986). As we shall see (Chapter 7), these conclusions basically reflect those derived from studies incorporating the genetics as well as the epidemiology of host–parasite interactions (May & Anderson, 1983).

Empirical evidence

The most basic assumptions of all models which investigate the regulatory effect of pathogens on plant populations are: (1) that pathogens reduce the competitive and/or long-term reproductive output of their hosts; (2) that a substantial proportion of the damage inflicted by pathogens is host-specific; and (3) that there is a density-dependent interaction between host and pathogen such that an increase in the numbers of a host species causes a disproportionate increase in the numbers of its pathogens.

The intricacies of host-specificity and of density-dependent interactions have been considered earlier (Chapter 4 and 3, respectively). It is sufficient to point out here that the pathology literature abounds with evidence concerning the host-specific nature of plant pathogens. Many pathogens are so highly specialized that individual species are subdivided into a complex series of sub-species (*forma speciales*) and races, each capable of attacking some and not other individuals within the pathogen's host range (e.g. the sub-specific and race structure of the cereal rust pathogen *Puccinia coronata* (Simons, 1970)). On the other hand, others, such as many of the species of *Phytophthora*, *Pythium* and *Rhizoctonia* responsible for seedling damping-off, have reputations for extremely wide host ranges. However, marked differences occur in the susceptibility of different hosts to these pathogens and, even within a host species, single gene changes may dramatically alter the susceptibility of individuals to infection (Bernard *et al.*, 1957). Similarly, there is clear evidence for the existence of density-dependent interactions between host and pathogen populations. Where such interactions have been studied in detail over the course of many pathogen generations, positive relations between the rate of increase of disease and the density of the host stands have been found for soil-borne and wind-borne diseases (Burdon & Chilvers, 1982).

The remaining assumption of ecological host–pathogen models, namely that pathogens may reduce the number, competitive ability and reproductive output of their hosts, needs to be considered in greater depth. Until recently almost all our knowledge of the effect that plant pathogens may have on their hosts has come from species of agricultural or forestry interest. In that context it is simple to demonstrate that, given favourable conditions, plant pathogens are capable of causing tremendous damage. This is seen in the classic disease epidemics of agriculture (e.g. potato blight in Ireland, 1847–50; southern corn leaf blight in North America, 1970). In contrast, because disease levels in non-agricultural systems rarely reach such heights, there has been a general tendency to ignore the effects of disease on the status of plant species in such situations. Fortunately, some evidence relevant to a consideration of the role of pathogens in plant ecology has accumulated from a variety of studies in natural and man-managed communities. While none of these studies provide even a modest testing of the theoretical models considered previously, they do indicate at least whether the broad assumptions of model builders are compatible with observations of the real world.

Single species populations

The effect of pathogens on population recruitment

One of the most intractable problems facing the population biologist is that of how to obtain records of the precise number of new recruits entering a population. At the germination and early establishment phase of life, plants, because of their small size, are particularly vulnerable to rapid elimination through the action of a range of abiotic and biotic factors. Newly emerged seedlings may quite literally be 'here today and gone tomorrow' and even the most conscientious monitoring may fail to record the transient existence of individuals which emerge and disappear again within a very short period of time. Moreover, the monitoring of seedlings arbitrarily defines 'birth' as seedling emergence and totally ignores the vast majority of seeds which enter the soil but never germinate. Up to 95% of seed that is shed from parent plants dies before emergence (Hickman, 1979). The inclusion of such individuals in the demographic analysis of populations inevitably results in a considerable change in our perception of the risks faced by individual plants during the course of the growth and development of populations. Survivorship curves change from those which show death risks to be concentrated late in the life cycle or to be relatively constant, to those which indicate massive early mortality (see Figure 1.1). Such changes have been observed for a range of species, both annual (e.g. *Minuartia uniflora* (Sharitz & McCormick, 1973)) and perennial (e.g. *Ranunculus* spp. (Sarukhan & Harper, 1973)).

Results of many studies of seedling recruitment have shown the effect of features of the microhabitat on the likelihood of successful establishment. The lack of sufficient soil moisture and the presence of seed predators and seedling herbivores appear to be the two most universally recognized sources of loss at this early stage. Paradoxically, however, changes in the micro-climate of potential germination sites which reduce the risk posed by such factors may dramatically increase that due to the activity of other, less obvious, biotic forces. Smith (1951), in a detailed study of the regeneration of white pine, found that while the risk of seedling death due to desiccation was substantially reduced in deep shade, this advantage was offset by an increase in the likelihood of fungal infection.

Pathogens characteristically associated with pre- and post-emergence damping-off (e.g. *Phytophthora*, *Pythium* and *Rhizoctonia* species) are readily recovered from many soils. Given suitable conditions these may have a substantial (>10% host mortality) effect on developing seedling populations (e.g. Bloomberg, 1973; King, 1977). The strong interaction between the level of activity of these organisms in the soil and environmental conditions (see 'the race to maturity', Chapter 2) inevitably results in marked patchiness of effect. This, combined with the inherent difficulties present in ascribing losses *a posteriori* to one of a range of factors, makes difficult an overall assessment of the role of soil-borne pathogens.

One study which has attempted to tackle this problem followed the individual fate of 440 Scandium-46-marked Douglas fir seeds released in a typical forest environment (Lawrence & Rediske, 1962). Although some seeds were also treated with bird and rodent deterrents, these treatments had no apparent effect. This allows the data to be re-organized to show the effect of pre- and post-emergence fungal attack relative to all other causes of loss (Table 6.1).

Mortality in this study was high, with 87% of all individuals succumbing during the first year – a result which is typical of virtually all populations studied in detail. The causes of death were divided between that attributable to fungi (43%) and those due to desiccation, predation and other undetermined factors. This study was particularly interesting as it docu-

Table 6.1 *Causes of death in a population of Douglas fir seeds planted in a forest environment. Data from Lawrence & Rediske (1962) have been reworked to show the importance of fungal damage*

Stage in life cycle	Causes of loss: Fungi	Causes of loss: All other	Cumulative: Loss	Cumulative: No. surviving
Pre-emergence				
(a) Pre-germination loss	86 (19.6)[a]	114 (25.9)	200	240 (54.6)
(b) Failed to germinate	40 (9.1)	78 (17.7)	318	122 (27.7)
Sub-total	126 (28.6)	192 (43.6)	–	– –
Post-emergence				
(c) Loss in first year	40 (9.1)	26 (5.9)	384	56 (12.7)
Total mortality	166 (37.7)	218 (49.6)	–	– –

[a]Percentages of original cohort are given in parentheses.

mented the greater mortality risk faced by seeds prior to emergence. During that stage 40% of deaths were caused by fungal action alone.

An alternative approach to the monitoring of pre-emergence death is the judicious use of fungicides to control selectively the depredations of pathogens. In agricultural trials involving the establishment of new pastures, fungicides have been applied either to seed prior to sowing (Kreitlow, Garber & Robinson, 1950; Dowling & Linscott, 1983) or directly to the soil (Clements *et al.*, 1982). Often, however, this has had remarkably little effect. The results of Michail & Carr (1966) are typical. For very small seeded species like the grasses cocksfoot and timothy, fungicide treatment prior to sowing resulted in a 15–20% better establishment over a 5 week period than in control plots (Figure 6.3). For larger seeded species like ryegrass, no difference in establishment rate was detectable. Moreover, even in those cases where fungicide treatment had an effect, the level of successful establishment never exceeded 50% of the initial viable seed population.

The reasons for such discrepancies have not been investigated. However, they may reflect the imposition of dormancy on sown seed, less than 100% efficiency of the fungicide utilized, or an enhancement of mortality due to micro-organisms like bacteria, favoured by the decline of fungal populations. Despite disappointing results to date with the use of fungicides, this

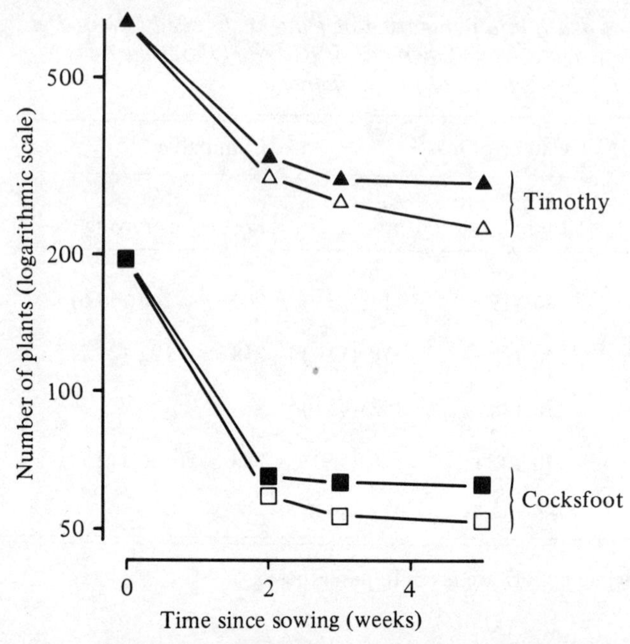

Figure 6.3 The establishment and survival of pasture grasses, cocksfoot and timothy, treated (●) and untreated (○) with a fungicide prior to sowing. (Data from Michail & Carr, 1966.)

approach still appears to offer some promise if used in properly designed experiments. A similar strategy has already been used successfully in plant–herbivore studies to show the importance of both slugs and insects (Foster, 1964; Cantlon, 1969).

After emergence, damping-off diseases may still seriously deplete the size of seedling populations (cf. the occurrence of disease in forestry seed-beds (Hartley, 1921)). Again, however, because affected individuals often disappear very rapidly and the ultimate cause of death may not be obvious, the *actual* contribution of such pathogens to seedling mortality is often not determined in field studies. One exception is a detailed examination of factors affecting the distribution and survival of seedlings of the neo-tropical tree *Platypodium elegans* (Augspurger, 1983*a*,*b*). In a study of the offspring of four trees, each isolated from the others, Augspurger found that a very high proportion of seedlings died during the first 3 months after emergence. The actual risk of death was strongly correlated with the distance of individual seedlings from their parent tree (negative correlation) and the density of seedling stands (positive correlation) (Augspurger & Kelly, 1984). For three of the four trees, 80% or more of the seedlings died, while around the fourth, 46% succumbed. Of all deaths, 72% were

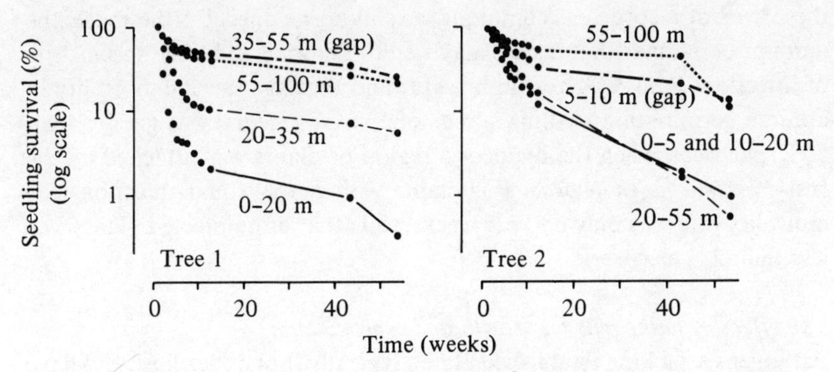

Figure 6.4 The percentage survival through time of seedlings of *Platypodium elegans* growing at various distances from two parent trees. The interactive effects of host, pathogen and environment on disease incidence is seen in the high survival of seedlings occurring in light gaps. (Redrawn from Augspurger, 1983*a*.)

caused by damping-off diseases alone; the remaining plants were killed by a motley array of biotic and abiotic forces.

Spatial analysis of the occurrence of damping-off showed that this decreased with distance from each parent tree. Seedlings close to their parents suffered higher mortality than those some distance away (Figure 6.4). Furthermore, the marked effect that environmental conditions may have on the interaction between plant and pathogen was clearly illustrated by the disproportionately high survival of seedlings growing in light gaps.

In a comparison of *Platypodium elegans* with seven other tree species occurring in the same area of tropical lowland forest, Augspurger (1984) found that her previous results were not typical of all species. The relative importance of fungal-induced seedling mortality varied greatly between the different species. For the six species *Cavanillesia platanifolia*, *Ceiba pentandra*, *Lonchocarpus pentaphyllus*, *Platypodium elegans*, *Tabebuia rosea* and *Terminalis oblonga*, pathogen attack was the largest single cause of early (< 2 months old) seedling mortality (ranging from 12 to 74%). Furthermore, this pathogen-induced mortality was much lower in light gaps than in adjacent shaded areas. However, the density-dependent and distance-dependent mortality relations found previously for *P. elegans* (Augspurger & Kelly, 1984) applied only to one other species – *L. pentaphyllus*. For the others, mortality levels were either similar regardless of the distance from the parent tree or, in the cases of *Aspidosperma cruenata* and *Triplaris cumingiana*, seedlings were largely unaffected by disease regardless of stand density or distance from their parent.

Seedling plants may also be attacked by air-borne pathogens, although these are not recorded as commonly as soil-borne ones. Furthermore, the amount of damage that occurs may well be determined by the specific site of infection. Paul & Ayres (in press) found that age-specific mortality in autumn germinating seedling stands of *Senecio vulgaris* was greater than 50% per week when the hypocotyl region of plants was attacked by the rust *Puccinia lagenophorae*. For plants with purely foliar infections the mortality rate was only 6% per week, while that of uninfected plants was less than 2% per week.

The effect of pathogens on growth and reproduction

Pathogens attacking established plants typically (but not exclusively) have their main effect through reductions in growth and general vigour. Ultimately, this affects fecundity, and, in the case of perennial species, longevity. Over time, even relatively minor changes in competitive vigour may produce considerable changes in the relative status of individual members of a population and their likely reproductive output. The inexorable effect of sub-lethal foliar infections can be illustrated by a simple competition experiment involving two genotypes of the composite weed *Chondrilla juncea*, one resistant and the other susceptible to a particular isolate of the rust fungus *Puccinia chondrillina*. In the absence of disease caused by this pathogen, these two genotypes have very similar competitive abilities. At the early rosette stage no differences were detectable, although by the time of flowering the susceptible genotype had a modest advantage (Burdon, Groves & Cullen, 1981; Burdon *et al.*, 1984). Size class frequency distributions in 1:1 mixtures showed an increasing dominance of individuals of the susceptible genotype with time (Figure 6.5a,b). By the beginning of flowering, all individuals greater than 1.2 g were of this genotype. The addition of the pathogen *P. chondrillina* to this system produced marked changes in the frequency distribution of individual plant weights (Figure 6.5c). Although at no time could one ascribe the death of susceptible individuals to the effects of the pathogen alone, its effect on competitive ability indirectly caused almost 60% of susceptible individuals to disappear (the difference between the number of susceptible individuals surviving in healthy and disease-infested mixed stands). In addition, the surviving susceptible individuals were all heavily suppressed.

These results for a perennial species are similar to those found when populations of the annual plant *Senecio vulgaris* are infected with *Puccinia lagenophorae*. In mixtures of infected and uninfected plants and also in stands in which all individuals were infected, Paul & Ayres (in press and

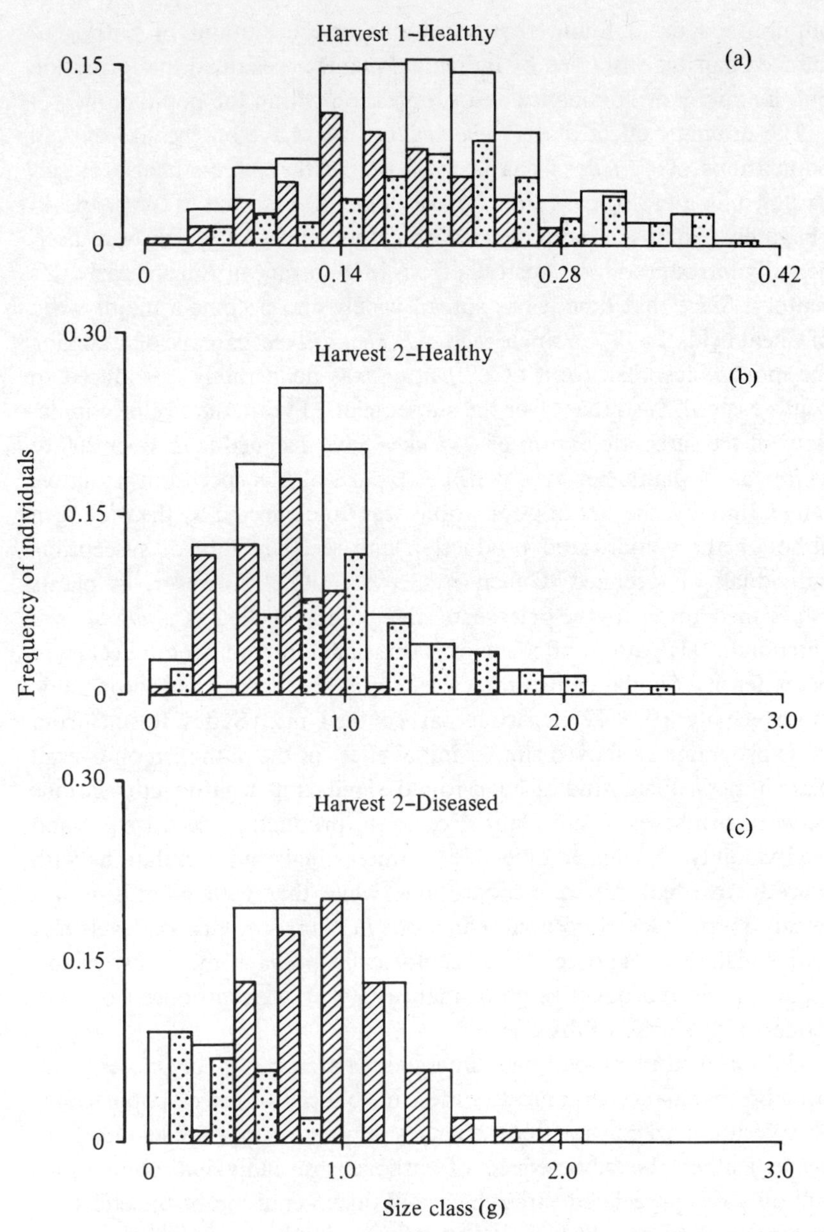

Figure 6.5 Size class distributions of two 1:1 mixtures of two genotypes of *Chondrilla juncea*, one susceptible and the other resistant to a particular isolate of the pathogen *Puccinia chondrillina*. (a) Distributions of susceptible (dotted) and resistant (hatched) genotypes at the rosette stage and prior to infection. (b), (c) Distributions at flowering in the absence and presence, respectively, of the pathogen. The size-class distribution for the population as a whole is given by the height of the unshaded region. (Redrawn from Burdon *et al.*, 1984.)

unpublished data) found that variations in the amount of pathogen-induced damage sustained by individual plants accelerated the formation of a hierarchy of dominance and suppression within the population.

The dramatic effect that *P. chondrillina* may have on the size of field populations of *C. juncea* has been shown in the successful use of this pathogen in a biological control programme in Australia. *C. juncea* is an obligately agamospermous species, three genotypes of which were accidentally introduced into Australia from Mediterranean Europe early this century. Since that time it has spread widely and become a major weed of wheat fields. In 1971 a single race of *P. chondrillina*, capable of attacking the most widespread form of *C. juncea*, was deliberately introduced to south-eastern Australia. Over the subsequent 15 years the size of populations of the susceptible form of *C. juncea* have declined from over 200 to as low as 10 plants per m^2 – densities typical of this species in its native range. Initially, the size of populations was little affected by the pathogen although the vigour, seed production and seed viability of susceptible individuals all declined (Cullen & Groves, 1977). However, as plants established prior to the release of the pathogen died (*C. juncea* is a perennial), this pattern of effects was translated into an overall decline in plant density. Similar deliberate releases of *P. chondrillina* have been made more recently (1978–79) in various parts of the United States. Results from these programmes show a similar initial effect of the pathogen on overall plant fitness. Field studies have found significant negative correlations between rust severity and plant size, flower production, seed weight and seed viability (Adams & Line, 1984). Interestingly, all correlations with reproductive features were logarithmic while that with plant size was linear. To produce noticeable reductions in plant size, disease levels had to be high. Less apparent, but ecologically perhaps more significant, reductions in reproductive performance (50%), were produced by very modest (5%) levels of disease.

Natural wild plant–pathogen interactions are presumably closer to an equilibrium balance than most of these introduced *C. juncea* populations. As a result, marked short-term changes in plant density are less likely to occur. Rather, the subtle effects of pathogens on individual populations will only be appreciated through careful documentation of the effects of disease on characters like longevity and reproductive performance. In one such assessment the effects of the pathogen *Synchytrium decipiens* on the annual legume *Amphicarpaea bracteata* was studied (Parker, 1986). A summary measure of the extent of early fungal infection was produced by estimating the number of lesions occurring on the most heavily infected sections of the first leaf and lower stem of host plants. In two separate populations of *A. bracteata*, pre-reproductive mortality rates were positively correlated with this estimate of early infection levels. Plants with disease levels above the median value were 3.8 to 12 times more likely to die before flowering than were lightly infected individuals. Furthermore, among those plants that reached reproductive maturity, disease levels were negatively correlated with total seed biomass.

Without doubt it is the general lack of easily recognizable relations between disease levels, survival and reproductive output (see Chapter 2) that has severely hampered the recognition of the potential role of pathogens during the vegetative and reproductive stages of the life cycle of many plants. This problem becomes particularly acute in the case of infections involving endophytic fungal pathogens that may be hidden for most of the host's life cycle. An example of these pathogens is *Epichloe typhina* (choke), which prevents flowering of its hosts. Here parasitic castration is compensated for by greater vegetative vigour so that infected plants may appear to be among the elite of a population although in reality they have reached an evolutionary dead end! Despite this, the vegetative vigour of infected plants can result in preferential survival in some habitats. By careful examination of individual plants of *Agrostis tenuis* and *A. stolonifera* sampled from 37 different populations. Bradshaw (1959) found that high levels of infection occurred in heavily grazed pastures. Very low or no infection was detected in situations that favoured reproduction by seed. It would appear that infection by *Epichloe* can result in individual plants that are more tolerant of a heavy grazing regime than are healthy ones!

In complete contrast to the examples given above, there are whole classes of pathogen whose effect on the reproductive output of their hosts is most obvious. One class of these is the diverse group of flower-infecting smut fungi. These organisms occur typically as systemic infections in their hosts and, at the time of flowering, replace the host's reproductive structures with a visually obvious mass of spores (Figure 6.6). In most host species, systemically infected individuals produce no seed at all. This makes the estimation of the effect of the pathogen on the reproductive fitness of the individual and its effect on the population as a whole, relatively simple.

Assessments of levels of smut infection in a range of hosts show tremendous variability both between different species and between different populations of the same species (Table 6.2). In natural populations of *Silene dioica* infected with *Ustilago violacea*, the level of infection

Figure 6.6 The effects of infection by the systemic floral smut *Sporisporium amphilophis* on *Bothriochloa macra*. Healthy plant on left, infected plant on right. The insets provide detailed pictures of individual healthy and infected inflorescences.

recorded in 20 separate populations ranged from zero to 32% of individuals. Interestingly, in this dioecious species the frequency of infection was often greater in female plants (but see Hassan & MacDonald, 1971). It has been suggested that although this disease does not apparently cause local extinctions it may be a long-term regulator of plant density (Lee, 1981). The same pathogen also causes smutting of *Viscaria vulgaris* in Sweden. There, the frequency of smutted plants was positively correlated with the size of host populations. Larger populations contained higher frequencies of the pathogen while small populations (less than 35 individuals) were free from disease. These observations led Jennersten and his associates (1983) to suggest that the pathogen was again acting as a regulatory mechanism

Table 6.2 *Representative examples of the mean level and variability of floral smut infection in a range of host species*

Host/pathogen		Number of populations examined	Mean infection (%)	Range of infection (%)	Reference
Silene alba/ *Ustilago violacea*		17	29.5	0–67.3	Baker, 1947
S. dioica/ *U. violacea*		20	11.0	0–31.8	Lee, 1981[a]
Viscaria vulgaris/ *U. violacea*	A[b]	25	15.8	0–55.8	Jennersten *et al.*, 1983
	B[b]	17	2.4	0–12.2	
Bothriochloa macra/ *Sporisporium amphilophis*		16	32.4	0–91.0	Burdon (unpublished data)

[a]Data are population means. Female plants in the same populations showed a mean infection of 15.2% and a range of 0 to 42.9%.
[b]Populations in two parts of Sweden.

Table 6.3 *The percentage of* Bromus tectorum *plants prevented from flowering by smut* (Ustilago bullata) *at three different sites in eastern Washington, United States, in the years 1977–80* (*data derived from Mack & Pyke, 1984*)

	Percentage smut infection in sites		
Year	Mesic	Moist	Dry
1977–78	42.6	30.8	2.1
1978–79	17.7	2.0	7.8
1979–80	25.3	2.0	36.6
3 year mean	22.7	3.1	10.3

slowing the increase in size of local *Viscaria* populations once they were sufficiently large to maintain the pathogen reliably.

A recent demographic analysis of three populations of *Bromus tectorum* infected with *Ustilago bullata* has shown just how variable the occurrence of smut infections may be from site to site and from year to year (Mack & Pyke, 1984). Over a 3 year period, smut infections accounted for between 2% and 43% of all 'deaths' (failures to reproduce) recorded in the three populations. Infections were generally more common in a population growing in a mesic environment than in populations growing in moist or dry situations (Table 6.3). However, even in the mesic environment, the proportion of plant mortality attributable to smut infection varied both between years and between different cohorts of individuals recruited within the one growing season. This variability in the occurrence of smut infection is not surprising, as the likelihood of successful infection is often strongly influenced by environmental conditions (Palti, 1981).

Even in these carefully documented populations, however, quantification of the effects of the smut is not as complete as might first appear. Many smuts, including *U. bullata*, initially infect their hosts at the seed or seedling stage, becoming obvious only at flowering. As a result, the recording of infections at this stage does not account for infected individuals that died during the vegetative phase of growth. In fact, in such populations we currently have no idea whether deaths which happen during the normal course of development (i.e. self-thinning) occur in a random sub-set of the stand or whether infected individuals are selectively eliminated. That the latter possibility may be the case is shown by studies of *Hordeum vulgare* infected with *Ustilago nuda* (Doling, 1964) and *Bromus catharticus* infected with *Ustilago bullata* (Falloon, 1976). In both instances, infected seedlings were found to be less vigorous than healthy ones.

Viruses and mycoplasmas are a second group of pathogens in which *some* infections cause generalized and noticeable effects on the whole plant. These organisms are widespread though not common in wild plant populations where, notwithstanding the previous comment, they usually cause few if any overt symptoms (Gibbs, 1983) (this is in distinct contrast to their effect on the majority of crop species). In some instances, however, infection is obvious. One such case is that of a mycoplasma-induced disease which causes stunting and the proliferation of chlorotic branches in *Erigeron canadensis*. In a population of *E. canadensis* growing on abandoned agricultural fields in Illinois, this mycoplasma produced symptoms of varying severity (Regehr & Bazzaz, 1979). Eleven per cent of the population was heavily infected and failed to produce any seed; a further 69% was either moderately or lightly infected. Respectively, these individuals produced only 12% and 44% of the mean number of seeds produced by healthy plants. Over the population as a whole, seed production was reduced by 53%.

Not all viruses which produce obvious symptoms in wild plants are totally detrimental. In one intriguing example, Gibbs (1980) has found that *Kennedia* yellow mosaic virus (KYMV), a tymovirus that produces bright symptoms in the young foliage of its wild host (*Kennedia rubicunda*), actually protects infected individuals against grazing by mammalian herbivores. Although virus infection decreased the growth rate of infected plants by about one third, in a natural community this was offset by preferential grazing. There, virus-free plants disappeared more than twice as quickly as infected ones (Figure 6.7). The tymoviruses as a group typically cause obvious symptoms even in wild hosts and Gibbs has suggested that many of these may also act in a manner similar to KYMV.

The previous example underlines some of the problems which arise when two or more 'parasitic' organisms act on a host population at the same time. Equally, in many circumstances fungal pathogens do *not* occur singly; rather, environmental conditions which favour one pathogen also favour others. Although the presence of these additional pathogens may complicate assessment greatly, estimation of the effect of only one pathogen, or even of each separately, may result in a severe underestimation of the overall effect of disease on the population. These problems are illustrated by a study of a population of *Capsella bursa-pastoris* in which the pathogens *Albugo candida* (white rust) and *Peronospora parasitica* (downy mildew) were both present (Alexander & Burdon, 1984).

In Minnesota, United States, both these pathogens overwinter in

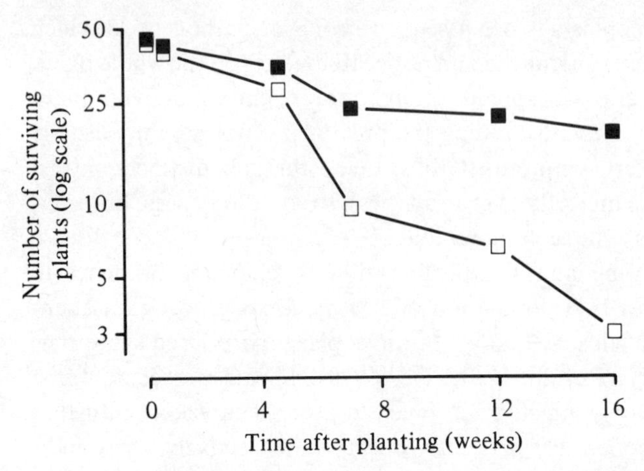

Figure 6.7 Survival of tymovirus (KYMV)-infected (■) and uninfected (□) young *Kennedia rubicunda* plants growing in natural bushland in the presence of herbivores, especially rabbits. (Redrawn from Gibbs, 1980.)

systemically infected seedlings and spread the following spring to uninfected plants, where they cause localized lesions. Individuals systemically infected with either pathogen were severely distorted (Figure 6.8), 90% dying before flowering. At a cursory glance, secondarily infected individuals appeared no different from healthy ones in their overall size, and the level of disease caused by each pathogen separately was not correlated with seed production. However, a combined disease rating detected a significant negative correlation between the level of disease caused by the two pathogens and reproductive performance. This study also showed the importance of the timing of disease occurrence. Plants infected earlier in the season were more affected than those attacked later, even though final disease levels were often the same.

Mixed populations

Random distribution of hosts

An earlier consideration of the theoretical aspects of plant–pathogen interactions showed that mixed stands greatly complicate the analysis of the effects of pathogens on their hosts. In disease-free situations, the performance of a species in a mixture may differ markedly from its behaviour when grown alone. In environments where diseases are present, the subtly different competitive interactions occurring between healthy individuals, between diseased individuals and between healthy and diseased individuals all contribute to a much more complex competitive

Figure 6.8 An individual plant of *Capsella bursa-pastoris* systemically infected with *Albugo candida* (white rust). Note the presence of pustules on leaves and stem over the entire plant. (From Alexander & Burdon, 1984.)

picture. Sorting out the relative effects of these different interactions is a necessary step in developing a thorough understanding of how pathogens may affect community composition. Unfortunately, however, work in this area is very much in its infancy.

The effect of mixed plantings of susceptible and resistant hosts on the size and rate of increase of pathogen populations has been studied extensively in agricultural situations. In the great majority of cases the severity of disease in mixtures is less (often markedly so) than that predicted by a simple proportionality of the level of disease occurring in monocultures of the component lines (for a full discussion see Chapter 3). However, from the plant's point of view, if these reductions in disease are to have any long-term ecological relevance they must be translated into positive changes in reproductive performance. What evidence is there for such changes?

Despite large numbers of assessments of disease occurrence in mixtures, relatively few studies have determined the effect that such disease levels have on reproductive output. Moreover, even these studies typically report only the total yield of mixed plots and pure stands planted with various component lines. As a result, although the data are sufficient to determine whether mixtures out-yield their component monocultures, they provide no basis from which to ascertain the cause of superior yields.

If the resource requirements of the component lines of a mixture do not overlap sufficiently, then even in disease-free situations mixtures will almost automatically out-yield the mean of their components. In mixed stands subject to pathogen attack, it is impossible to distinguish this possibility from a true disease reduction effect unless disease-free controls are grown. In the results of 74 two- and three-component mixtures culled from the literature (Table 6.4), only those of Jeger, Jones & Griffiths (1981) recorded such controls. Moreover, in their experiment involving two barley varieties and *Rhynchosporium secalis* (scald), the yield increase which could be attributed to a disease reduction effect was only +1.8% (Table 6.4; 4.1 to 2.3%). Similarly, in their experiment involving two wheat varieties and *Septoria nodorum* (glume blotch) an apparent mixture advantage in the presence of disease of +17.4% was totally explained by an over-yielding effect due to the same mixture in the absence of disease of +19.4% ! These data are sufficient to underline the need for caution in interpreting the results of many mixture experiments as necessarily reflecting a yield advantage linked to disease reduction. However, data of the type presented by Wolfe & Barrett (1980; Table 6.4) indicates that the effectiveness of mixtures varies with the severity of damage. Overall, the disease controlling effect of 47 three-component mixtures resulted in a yield increase of +6.5%. At the ten sites at which disease was most severe the average yield advantage rose to +9.0%.

Even when the effects of differential resource requirements are elimi-

Table 6.4 *The distribution of seed yields of intra-specific mixtures compared with yields of their component monocultures in the presence of disease*

Host/pathogen combination	Mixture yield[a]				Percentage advantage over P_M	Reference
	P_L	P_L to P_M	P_M to P_H	P_H		
Equal proportion two-component mixtures						
Avena/Helminthosporium victoriae	–	–	1	1	+9.0	Ayanru & Browning, 1977
Hordeum/Erysiphe graminis	1	–	–	6	+6.3	White, 1982[b]
Hordeum/Rhynchosporium secalis	–	–	1	–	+4.1	Jeger *et al.*, 1981[c]
Phaseolus/Sclerotinia sclerotiorum	–	–	3	1	+8.3	Coyne *et al.* 1978
Triticum/Erysiphe graminis	–	–	1	–	+7.6	Fried *et al.*, 1981
Triticum/Puccinia recondita	–	–	1	–	+3.1	Klages, 1936
Triticum/Puccinia striiformis	–	3	1	–	−1.3	Groenewegen & Zadoks, 1979
Triticum/Septoria nodorum	–	–	1	–	+17.4	Jeger *et al.*, 1981[d]
Equal proportion three-component mixtures						
Hordeum/Erysiphe graminis	–	–	1	5	+9.0	Day, 1981
Hordeum/Erysiphe graminis	–	8	13	26	+6.5	Wolfe & Barrett, 1980[e]

[a] P_L and P_H are the yields of the highest and lowest yielding monocultures, respectively; P_M is the mid-monoculture yield.
[b] 1980 data only. [c] Disease-free control mixtures yielded 2.3% above P_M. [d] Disease-free control mixtures yielded 19.4% above P_M.
[e] If lightly diseased mixtures are excluded, average yield was 9.0% above P_M.

nated, over-yielding ($P_M > (P_L + P_H)/2$) may still occur in disease-affected situations. In these circumstances, two explanations are possible. First, susceptible members of mixed stands are less damaged than those in pure stands. Secondly, resistant individuals will show compensatory yield increases (the extent of this will depend on the timing and severity of pathogen attack; late effects leave little time for compensatory responses by resistant individuals (Hirst *et al.*, 1973; Ross, 1983)). In real mixtures both of these features are likely to occur. To determine their relative importance, total mixture yields have to be broken down into their component fractions. Unfortunately, because this level of detail has usually been overlooked by plant breeders and pathologists interested in mixtures for reasons of disease control alone, data of this type are not available. For population biologists, on the other hand, such information is a vital step in assessing the likely long-term outcome of competitive interactions in the presence of disease.

In Chapter 7 two examples of temporal fluctuations in the relative frequency of different plant genotypes in disease-affected mixtures are discussed. These examples could equally have been addressed here; however, because they deal with changes in the frequency of genotypes of just the one species they were judged particularly relevant to a consideration of the effects of pathogens on the genetic structure of plant populations. In the context of the present chapter, it is appropriate to examine two other experimental systems that give some insight into the way in which pathogens affect not only their hosts but, indirectly, other plant species in the same community. Unfortunately, both these studies are restricted to changes occurring during a single generation.

In a comparison of the competitive relationships occurring in mixed stands of barley and wheat, Burdon & Chilvers (1977*b*) calculated the relative rate at which barley replaced wheat in a 1:1 mixture over the course of five consecutive vegetative harvests (the 'relative replacement rate' of van den Bergh & Ennik (1973)). In the absence of disease, barley became more and more dominant in the mixture as time passed (Figure 6.9a). However, in the presence of the barley-specific pathogen *Erysiphe graminis hordei* (powdery mildew), the performance of individual barley plants was reduced and the relative contribution of the two species to the mixture was approximately equal. If these pathogen-induced changes in competitive ability were to occur over a number of generations they could provide stability to highly biased competitive interactions.

In contrast, the results of an experiment devised by Groves & Williams (1975) provide a clear indication of the consequences of pathogen damage to the less competitive of a pair of cohabiting plant species. In the absence

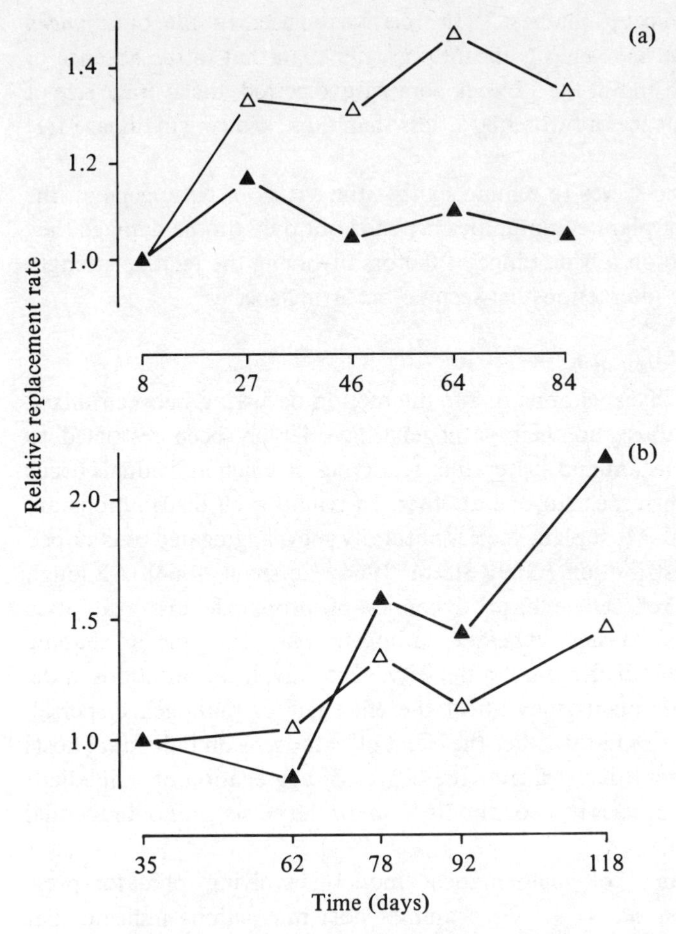

Figure 6.9 Two contrasting results showing the effects of differential pathogen attack in mixed plant stands. (a) The relative replacement rate of barley with respect to wheat in the presence (▲) and absence (△) of mildew of barley caused by *Erysiphe graminis*. (Redrawn from Burdon & Chilvers, 1977*b*.) (b) The relative replacement rate of subterranean clover with respect to skeleton weed (*Chondrilla juncea*) in the presence (▲) and absence (△) of *Puccinia chondrillina* (rust). (Data from Groves & Williams, 1975.)

of disease *Trifolium subterraneum* normally dominates mixtures with *Chondrilla juncea*. As time progresses, the flat, rosette leaves of the latter species become increasingly shaded by the higher canopy of clover leaves (Figure 6.9*b*). This effect may be so intense that after 17 weeks the average size of *C. juncea* plants growing with *T. subterraneum* is only 30% of that when grown alone. The presence of the rust pathogen *Puccinia chondrillina* further exacerbates the poor competitive performance of its host. After an initial minor 'hiccup' (harvest 2), the relative replacement rate of *C. juncea* by *T. subterraneum* was significantly greater than that in the absence of disease. By the end of the 17 week competitive period, the average size of *C. juncea* plants was a further 80% less than those growing in disease-free mixtures!

This example serves to remind us that the effects of pathogens on the composition of plant communities depends not only on the damage they cause, but also on a wide range of factors involving the relative strength of competitive interactions between co-occurring species.

Patchy distribution of hosts

Experimental investigations of the interaction occurring between mixed plant populations and their pathogens have largely been restricted to mixtures of resistant and susceptible genotypes in which individuals occur randomly with respect to one another. In reality such distributions are highly artificial. Most plant species naturally show aggregated or clumped patterns of distribution (Greig-Smith, 1964; Kershaw, 1964). Although these largely reflect the basic dynamics of propagule dispersal, it is appropriate to consider whether pathogens play any role in shaping patterns of plant distribution in the field. Certainly, local variations in the pattern of host plants may affect the efficiency of pathogen dispersal. However, can this in turn alter the effect of pathogens on individual hosts to such an extent that it affects the degree of aggregation of individuals of a particular species in a community and/or the persistence of individual patches?

A wide range of mathematical models involving predator–prey, parasitoid–host or even disease–animal host interactions indicate that environmental heterogeneity can make a substantial contribution towards the stability of individual host–consumer interactions (Vandermeer, 1973; Roff, 1974; Hilborn, 1975; Hastings, 1977; Anderson & May, 1978; Hassell, 1980). By restricting the distribution of the host population to a number of discrete patches, between which both host and consumer may disperse, the probability of extinction of either population as a whole is reduced substantially below that likely in a single homogeneous environment. Unfortunately, as pointed out at the beginning of this chapter, sophisticated models of this type have yet to be developed for the interaction occurring between plants and their pathogens. However, the basic proposition that aggregated patterns of host and pathogen distribution may lead to greater stability seems intuitively reasonable. Thus, in line with the theory of island biogeography (MacArthur & Wilson, 1967), if patches of hosts are seen as host 'islands' growing in a 'sea' of non-hosts, then small patches (or islands) that are well separated from the main body of individuals may: (1) escape initial infection for longer; (2) fail to support the continued existence of pathogens from season to season; and (3) at any given time sustain fewer pathogen individuals. The merits of such an argument rest on the relative success of the dispersal of pathogen propagules within and between host patches (see auto- and allo-infection, Chapter 3).

In agriculture this question is reflected in the controversy which has arisen over the use of small or large fields (van der Plank, 1948, 1949, 1960; Waggoner, 1962). That is, are severe disease epidemics more likely to occur in crops that are concentrated into a few large but widely spaced fields or in those grown in many smaller but less distant fields? By placing different emphases on the probability that pathogen propagules released from one field would give rise to infections in another, and on the relative efficiency of within- and between-field dispersal, van der Plank and Waggoner came to quite different conclusions. van der Plank (1948, 1949, 1960) reasoned that making fields uniformly larger and correspondingly fewer and further apart, decreased the likelihood of a severe disease epidemic. The opposite view was espoused by Waggoner (1962) who argued that a severe epidemic was less likely to occur if fields were small and scattered. Further theoretical support for this latter view has come from a simulation model of epidemic development (Zadoks & Kampmeijer, 1977) and a complex mathematical model involving time-related changes in inoculum production and dispersal (Fleming, Marsh & Tuckwell, 1982). In both of these studies, when a constant proportion of a region's acreage was allocated to a crop, decreasing field sizes were generally found to retard disease development. However, Zadoks & Kampmeijer (1977) did find that, ultimately, specific solutions were largely determined by the frequency, size and distribution of susceptible fields, by the number, position and size of sources of inoculum, and by factors associated with its dispersal.

For two very different plant diseases, crown rust of oats and damping-off of cress, there is some experimental evidence showing that the development of pathogen populations may be affected by host plant distribution. In the former case, Mundt and his colleagues (Mundt & Browning, 1985; Mundt & Leonard, 1985) have examined the effects of grouping like genotypes in mixtures of susceptible and resistant oat plants on the ability of the mixture to reduce the rate of development of epidemics of *Puccinia coronata* (crown rust). When epidemics developed from single foci of infection, disease levels were substantially lower in all the patterned

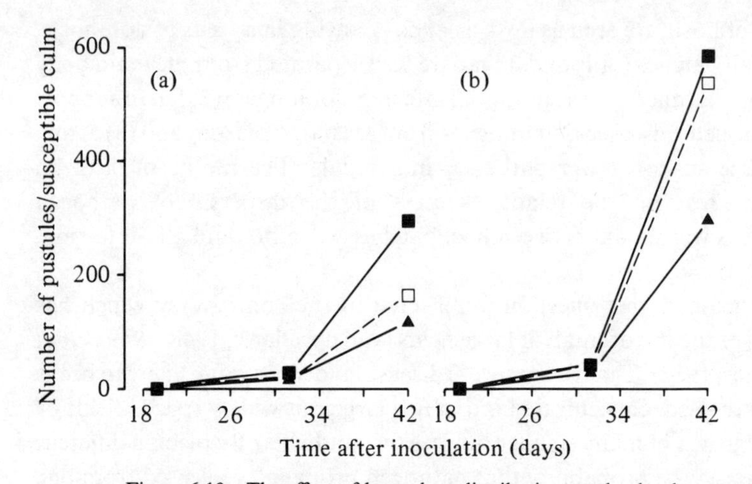

Figure 6.10 The effect of host plant distribution on the development of epidemics of *Puccinia coronata* (crown rust) in oats. (a) Epidemics were initiated by artificial inoculation of a central focus in each plot. (b) Epidemics were initiated by distributing inoculum uniformly over the entire plot. ■, disease progress curve for a pure stand of the susceptible oats; □, disease progress curve for a 3:1 mixture of immune and susceptible oats in which seeds were aggregated into blocks of 200 seeds of like genotype; ▲, disease progress curve for a completely random 3:1 mixture of immune and susceptible oats. Each plot is the mean of three replications. (Redrawn from Mundt & Leonard, 1985.)

mixtures than those occurring in pure stands of susceptible individuals. In some experiments, no significant differences were detected in disease development, even when the areas occupied by distinct patches of susceptible plants ranged from 30 to 8400 cm² (Mundt & Browning, 1985). In a similar study, Mundt & Leonard (1985) also detected no significant difference between the development of disease in a random 3:1 mixture of resistant and susceptible individuals and another mixture in which the plants were grouped into patches of 5800 cm² containing 200 plants of the same genotype (Figure 6.10). In both of these mixtures the level of disease development was markedly lower than that measured in a pure stand of the susceptible component. Importantly though, Mundt & Leonard (1985) showed that if an epidemic developed as a result of a general distribution of inoculum across an area (as would occur if the inoculum source was some distance away, see Chapter 3), then the efficacy of patterned host mixtures for reducing disease was eliminated (Figure 6.10).

The effect of the number and distribution of initial infection foci on the rate of development of disease epidemics in aggregated plant stands has also been shown in a study involving epidemics of damping-off (induced by *Pythium irregulare*) in patterned stands of *Lepidium sativum* (Burdon & Chilvers, 1976*c*). In experiments in which disease developed from a single focus, the rate of linear advance of damping-off was unaffected by changes in the size of individual patches, provided that the total area occupied by these remained constant. However, if changes in disease levels were monitored in a system involving the development of multiple infection foci from a random background inoculation, the rate of disease development rose sharply when the size of individual host patches was large. Individual patches that were large generally contained at least one focus of infection. On the other hand, host patches that were small often escaped initial infection and became diseased only after secondary spread of the pathogen from adjacent infected patches. Burdon & Chilvers (1976*c*) also examined the effect of altering the total area occupied by patches. Not surprisingly, both the rate of linear advance and the overall rate of increase of damping-off were inversely related to the distance between adjacent patches. Although the proportion of patches containing primary infection foci was independent of the overall area occupied, the much greater distance between adjacent patches reduced the efficiency of between patch transmission and hence reduced the rate of disease development.

Results like these confirm that there are at least some circumstances under which disease development may be enhanced by some patterns of host plant distribution and retarded by others. As yet, little evidence is available to determine whether these differences last long enough to be translated into differences in the fecundity and survival of individuals occurring in different patches in natural situations. However, Jennersten *et al.* (1983) found a significant positive correlation between the size of patches of *Viscaria vulgaris* and the occurrence of infection by *Ustilago violacea* (smut), a pathogen that causes permanent sterility of its host. As patch size increased, the frequency of permanently infected plants rose and thus, by inference, the long-term viability of the patch declined. In a similar way it has been suggested (Pratt, Heather & Shepherd, 1973) that the restriction of many eucalypt species to ridge-top sites in south-eastern Australia is caused by the extreme susceptibility of their seedlings to attack by pythiaceous fungi like *Phytophthora cinnamomi* (root rot). Ridge-top sites represent safe refuges for susceptible species because the drier soil conditions that occur there inhibit pathogen development and spread.

Communities

Ultimately, most plant species do not occur in monospecific stands spread over large areas or even in simple two- or three-component mixtures. Rather, they occur in complex communities in which the loss or debilitation of individuals of one species may have far-reaching effects on a large number of associated ones. Studies of both the primary (on hosts) and secondary (on non-hosts) effects of diseases in whole communities are rare and are generally associated with serious diseases of economically important tree species. Such diseases may be caused by either introduced or native pathogens, although in both cases recent human activities seem to be a vital link in the chain of events.

Probably the best known and documented example of an introduced pathogen that devastated natural populations of a previously unexposed host is that of *Endothia parasitica* (chestnut blight) and its host, the American chestnut (*Castanea dentata*). In its home range of eastern Asia, this pathogen causes relatively little damage to the native species of chestnut. When introduced into North America, however, the pathogen almost completely eliminated its new host from large areas of forest. Day & Monk (1974) found that the basal area of *C. dentata* in a mixed stand declined from 31.1% in 1950 to 0.1% in 1970. This was accompanied by a rise in importance of *Quercus prinus*, *Acer rubrum*, *Liriodendron tulipifera* and various species of hickory (*Carya* spp.). Nelson (1955) and Woods & Shanks (1957) also found similar tendencies for a range of existing dominant and co-dominant species to benefit from the decline in size and density of the chestnut population.

On a number of occasions native pathogens have also been found to cause extensive changes to communities dominated by their hosts. The finely balanced nature of many of these natural host–pathogen interactions is well illustrated by the rise in prominence of fusiform rust (caused by *Cronartium fusiforme*) of southern pines since the advent of European man to North America. The pine forests of the southern United States are a fire sub-climax in which, in the absence of fire, succession proceeds towards a mixed hardwood climax (Dinus, 1974). Prior to widespread disturbance by European man, fires maintained and determined the composition and structure of the pine forests so that while succession was checked, hardwoods, including oaks, were present in the understorey. However, despite this natural association of pines and oaks (*C. fusiforme* is an obligately heteroecious rust alternating its life cycle between pine and oak hosts) fusiform rust was rare (Czabator, 1971).

Agricultural and forestry activities of man since the 1870s have dramati-

Figure 6.11 Changes in the appearance of an Australian sclerophyll shrubby woodland associated with the presence of the pathogen *Phytophthora cinnamomi*. (a) Vegetation of a disease-free control plot; (b) vegetation of an infected plot. (From Weste, 1981.)

cally altered this balance by contributing to an increase in the distribution and density of susceptible oak and pine species and to a decrease in the occurrence of resistant species. This has occurred as a result of: (1) the early selective felling of resistant species of pine; (2) the clearing of land for farming; (3) the land's abandonment and subsequent colonization by susceptible pine species; (4) the widespread commercial planting of susceptible species; and (5) the exclusion of wild fires. These changes in forest composition and structure have altered the natural balance of this host and pathogen system to such an extent that an epidemic of lasting significance has resulted (Dinus, 1974).

The second case to be considered here concerns the spread of *Phytophthora cinnamomi* (root rot) into a range of vegetation types in Australia. Unlike the pathogens of the previous examples *P. cinnamomi* has an extremely wide host range attacking plants from at least 48 different families (Newhook & Podger, 1972). It appears to be endemic to eastern Australia (Heather, Pratt & Shepherd, 1975) but was recently introduced to Western Australia where the pathogen is inexorably spreading through large areas of eucalypt forest (Podger, 1972). Even in the east, where it is endemic, localized epidemic outbreaks still cause drastic changes to the flora. Weste (1981) has documented these changes by comparing, within a eucalypt shrubby woodland, plots established in an area infested with the pathogen with ones in an adjoining disease-free area (Figure 6.11). Even at the beginning of the study considerable differences were found between the two sites in the occurrence of individuals of a wide range of tree and shrub species (Table 6.5). Presumably these differences reflected the past activity of the pathogen. Over the following 5 years changes in the relative frequencies of different species continued so that by 1979 a complete change was seen in the structure and composition of the plant community. The vegetation of the infected plot changed from a sclerophyll woodland with a thick *Phytophthora*-susceptible shrub understorey to an open woodland dominated by the resistant sedge *Lepidosperma concavum*. More than 70% of plants in the infected plot were of this species as opposed to less than 20% in the healthy stand. Subsequent visual examination of the plots in 1980 indicated that all the remaining *Banksia serrata* trees in the diseased area were infected. When these die, an open sedge community will remain.

Similar changes, although of varying severity, have been recorded in other Australian plant communities on different soil types (Weste, 1974, 1980). Whether the pathogen will disappear from these communities with the loss of susceptible hosts or whether sufficient regeneration of

Table 6.5 *Vegetational changes associated with* Phytophthora cinnamomi-*induced die-back in a south-eastern Australian eucalypt woodland (from Weste, 1981)*

	Number of individuals recorded in				Diseased plot numbers as percentage of healthy	
	Diseased plot		Healthy plot			
Species[a]	1974	1979	1974	1979	1974	1979
Trees						
Banksia serrata (S)	64	22	46	43	139.1	51.2
Eucalyptus obliqua (S)	8	3	60	5	13.3	60.0
Shrubs						
Acrotriche serrulata (S)	2	2	17	5	11.8	40.0
Bossiaea cinerea (S)	698	533	470	776	148.5	68.7
Dillwynia sericea (S)	330	124	517	766	63.8	16.2
Epacris impressa (S)	65	20	695	714	9.4	2.8
Hakea sericea (S)	17	11	8	13	212.5	84.6
Hibbertia virgata (S)	1	4	73	19	1.4	21.1
Isopogon ceratophyllus (S)	0	0	59	37	0	0
Leptospermum myrsinoides (S)	113	35	95	122	119.0	28.7
Leucopogon virgatus (S)	306	10	210	40	145.7	25.0
Tetratheca ciliata (S)	43	13	630	160	6.8	8.1
Xanthorrhoea australis (S)	47	17	644	597	7.3	2.9
Acacia suaveolens (T)	162	226	38	62	426.3	364.5
Casuarina pusilla (T)	124	95	107	82	115.9	115.9
Hypolaena fastigiata (R)[b]	30	50	5	5	600.0	1000.0
Lepidosperma concavum (R)	4186	3089	1674	780	250.1	396.0

[a](S) = susceptible, (T) = tolerant and (R) = resistant to disease. [b]Values recorded for *H. fastigiata* are per cent cover as this species spreads by means of underground rhizomes.

susceptibles will occur to maintain the local pathogen population is unknown. However, declines in pathogen disease potential following disease episodes have been observed (Weste, Cook & Taylor, 1973) and Weste (1981) has suggested that this pathogen and its hosts may respond to one another in a cyclical fashion.

7

The effect of pathogens on the genetic structure of plant populations

Having previously examined the potential and actual roles of pathogens as determinants of the size of individual plant populations (Chapter 6), we here turn to the question of under which circumstances the reductions in survival and reproductive capabilities induced by pathogens lead to changes in the genetic structure of host populations. What evidence is there for pathogen-mediated changes in the genetic structure of real plant populations? How do such populations respond to changes in the selection pressure applied by pathogens? The present chapter addresses these questions and attempts to provide some answers.

The basic co-evolutionary scenario

All models concerned with the interaction of host and pathogen populations at a genetic level are founded on the same basic argument. This recognizes the inherent advantage that a novel resistance allele confers in an otherwise uniformly susceptible host population confronted by a uniformly avirulent pathogen population (note that this combination of host and pathogen genotypes results in susceptible infection type responses!; see Chapter 4).

At first, because of the low frequency that such a novel resistance gene has in the host population, selective pressure on the pathogen population favouring the emergence of the complementary virulence gene will be negligible. However, as the advantage of the mutant host genotype is translated into an increasing frequency of resistant individuals, this provides more scope for the appearance of a chance mutation in the pathogen population that possesses an appropriate virulence gene. When this occurs, the selective advantage previously enjoyed by the mutant host genotype is largely lost and the interaction between host and pathogen completes the first twist of an apparently unending co-evolutionary spiral. So long as the pathogen continues to exert strong selective pressure on the host, a continuation of this interactive process will lead to more and more diversified populations. Eventually there will be many resistance genes present in the host population and many virulence genes in that of the pathogen. Any particular resistance gene will provide protection against only a sub-set of the pathogen population, while any individual pathogen race will be virulent on only some of the host population. Person (1966) proposed that a balance would finally be achieved between the two populations through a series of cyclic polymorphisms between specific resistance alleles in the host and complementary virulence characters in the pathogen.

Theoretical models

Gene-for-gene population models

Host–pathogen mathematical models that analyse various aspects and implications of the preceding heuristic argument can be divided broadly into two groups: those in which the behaviour of both host and pathogen is of interest, each being allowed to respond to changes in the other; and those in which the behaviour of the pathogen population is the centre of attention, the size and structure of the host population being held constant. Most of these models specifically address situations involving host plants and fungal pathogens and are based on the interaction of resistance and virulence characters which conform to the gene-for-gene theory (see Chapter 4). This does not imply, however, that broadly based quantitative resistance in the host or aggressiveness in the pathogen is not important in contributing to the long-term outcome of such interactions in the real world. Rather, because less is known about genetic interactions between hosts and pathogens with respect to these characters, little attempt has been made to analyse their interaction mathematically.

Turning then to population models in which host and pathogen interact over time scales which permit both to reproduce, we can address questions concerning the long-term consequences of such interactions. It is not intended to consider all such models in detail here. Instead, two of the most appropriate (Jayakar, 1970; Leonard, 1977) will be used to illustrate the problems encountered and the basic biological assumptions which such models must make in order to produce conditions under which polymorphisms, between resistance and susceptibility in the host and virulence and avirulence in the pathogen, may ensue.

Jayakar (1970) advanced a simple host–parasite (pathogen) model* in which he envisaged that: (1) infection of a susceptible host by a pathogen

* Where appropriate the nomenclature used in the various models discussed here has been standardized for ease of comparison. As a result, symbols sometimes differ from those used in the original references.

Table 7.1 *The number of surviving host and pathogen progeny after one generation in the theoretical model developed by Jayakar (1970) (from Leonard & Czochor, 1980 and reproduced, with permission from the* Annual Review of Phytopathology, *Vol.* **18**. *© 1980 by Annual Reviews Inc.)*

Pathogen genotype	Host genotype r (susceptible)	R (resistant)
(a) Host numbers		
V (avirulent)	$Np_1p_2(1-x)$	$N(1-p_1)p_2$
v (virulent)	$Np_1(1-p_2)(1-x)$	$N(1-p_1)(1-p_2)(1-x)$
(b) Pathogen numbers		
V (avirulent)	np_1p_2x	0
v (virulent)	$np_1(1-p_2)x$	$n(1-p_1)(1-p_2)x$

p_1 = frequency of susceptible host individuals; p_2 = frequency of avirulent pathogen individuals; x = probability that a host individual will be affected if it is susceptible to the pathogen genotype encountered; N = number of progeny produced by healthy host individuals; n = number of pathogen progeny produced in an infected host.

results in the death of that host, while the pathogen produces an average of n offspring; (2) uninfected hosts produce an average of N offspring; (3) the probability of infection of any particular susceptible host is x; (4) the interaction involves a single diallelic locus in a haploid host (R,r) and pathogen (v, V) behaving in a standard genetic fashion (see Chapter 4); and (5) the frequencies of the four genotypes r, R, V and v are p_1,$1-p_1$, p_2 and $1-p_2$, respectively. Using this framework, Jayakar (1970) showed that the proportion of susceptible (r) hosts infected by avirulent (V) pathogens is xp_1p_2, with the consequent production of nxp_1p_2 pathogens in the next generation. While all such infected host individuals fail to reproduce, the remaining $(1-x)$ susceptible, but uninfected, individuals in this combination produce $Np_1p_2(1-x)$ offspring. Similarly, because the proportion of avirulent pathogens (V) that encounter resistant hosts (R) is simply given by $(1-p_1)p_2$, and as these hosts do not become infected, pathogen individuals in this combination fail to reproduce while the hosts reproduce and contribute a total of $N(1-p_1)p_2$ offspring to the next generation. The numbers of surviving offspring of all genotypes of host and pathogen after one generation are summarized in Table 7.1.

The only biologically meaningful equilibrium which results from this model occurs when $p_2 = 0$. That is where the virulence (v) allele becomes fixed in the pathogen population and consequently the entire host

Table 7.2 *The fitness of host and pathogen genotypes in the theoretical model developed by Leonard (1977) (from Leonard & Czochor, 1980 and reproduced, with permission, from the* Annual Review of Phytopathology, *Vol.* **18**. *© 1980 Annual Reviews Inc.)*

Pathogen genotype	Host genotype *rr* (susceptible)	 *Rr/RR* (resistant)
(a) Fitness of pathogen genotypes		
V (avirulent)	1	$1-t$
v (virulent)	$1-k$	$1-k+a$
(b) Fitness of host genotypes		
V (avirulent)	$1-s$	$1-c-s(1-t)$
v (virulent)	$1-s(1-k)$	$1-c-s(1-k+a)$

k = cost of virulence; t = effectiveness of resistance in suppressing pathogen reproduction; a = advantage to the pathogen of having its virulence gene match the corresponding resistance gene in the host; s = suitability of environment for disease development expressed in terms of loss of fitness of the susceptible host when attacked by the avirulent pathogen genotype; c = cost of resistance.

population becomes vulnerable, regardless of the presence or absence of the resistance allele (*R*). In these circumstances, the host alleles (*r*,*R*) are selectively neutral.

A balanced polymorphism of the type suggested by the earlier heuristic argument only becomes possible through the addition of selection which counterbalances the inevitable tendency for the virulent pathogen genotype to dominate. Jayakar (1970) achieved this by introducing simple fitness differences for genotypes of both host and pathogen. Specifically, if it is assumed that the fitness of resistant individuals relative to that of susceptible ones is f and that of virulent pathogens relative to avirulent ones is g, then a non-trivial equilibrium point exists at $p_1 = g$ and $p_2 = (1-x)\,(1-f)/xf$ when $g < 1$ and $(1-x) < f < 1$. In contrast to the earlier model in which the virulence gene became fixed in the pathogen population, here host and pathogen gene frequencies cycle around a locally unstable equilibrium point in a closed ellipse, so that a balanced polymorphism is achieved (Jayakar, 1970; Leonard & Czochor, 1980). Even if this model is complicated by conversion of the host from a haploid to a diploid state, a similar non-trivial equilibrium point occurs.

Leonard (1969*b*, 1977) developed independently a model involving the interaction occurring at a single diallelic locus in a diploid host and a haploid pathogen. He also found that unless the fitness of host and pathogen genotypes differed according to the particular combination in which they occurred, then the frequency of the virulence gene rapidly approached unity and a balanced polymorphism was not sustained. To overcome this, Leonard assigned a complex series of fitness values to the various host genotype–pathogen genotype combinations (Table 7.2). Thus, taking the fitness of the avirulent pathogen genotype attacking a susceptible host as 1, the fitness of this genotype on a resistant host is reduced by a value t. This represents the effectiveness of the resistance gene in suppressing pathogen sporulation (resistance in this diploid host model is assumed to be completely dominant, so that genotypes *Rr* and *RR* have exactly the same phenotypic response). Clearly, the value of t may vary with different resistance genes, but in situations where they produce major phenotypic effects (typically, those for which these models have been developed), $t > 0.85$ (Leonard, 1969*b*, 1977; Marshall & Burdon, 1981).

In the case of the virulent pathogen genotype confronting a susceptible host, the fitness of the pathogen is reduced from that in an avirulent pathogen–susceptible host combination by a value of k. This represents the cost of virulence. Finally, the value a (virulent pathogen–resistant host combination) allows for the possibility that the virulent pathogen genotype may reproduce equally effectively on susceptible and resistant hosts ($a = 0$) or be more effective on resistant hosts ($a > 0$). Similar fitness values have been proposed in another model developed to examine changes in the pathogen population alone (Person, Groth & Mylyk, 1976; Groth & Person, 1977).

The fitnesses of the two host genotypes were developed in a similar manner with the term s (Table 7.2) being a measure of the severity of disease. As s approaches 1, disease severity increases. For resistant genotypes, the term c represents the cost of this resistance compared to the fitness of susceptible individuals in a disease-free environment. Finally, the actual fitnesses of individual host genotypes are obtained by incorporating the fitness of the individual pathogen genotypes infecting these hosts (Table 7.2; Leonard, 1977). These values are then used to estimate the relative fitness of the host and pathogen genotypes in the presence of mixed host and pathogen populations.

Mathematical analysis of the model derived from this framework shows the existence of eight trivial and one non-trivial equilibrium points (Leonard & Czochor, 1980). The latter point occurs at $F_R = k/(a+t)$ and $F_V = (ts-c)/(ts-as)$, where F_R and F_V are the frequencies of resistant host genotypes (*RR* and *Rr*) and of the virulent pathogen genotype, respectively. This non-trivial equilibrium point is not stable (Leonard & Czochor, 1978, 1980; Sedcole, 1978). Computer simulation shows that when gene frequencies are not close to extinction or fixation they will cycle around the point in a closed ellipse. On the other hand, when initial gene frequencies are close to either extreme limit they spiral inwards towards the equilibrium until, presumably, the amplitude of the oscillations in gene frequencies equals that of an uncharacterized limit cycle (Leonard & Czochor, 1978, 1980; Fleming, 1980).

In mixed host–pathogen interactions, the fitness of each host genotype in both the Jayakar (1970) and Leonard (1977) models, is the mean of its fitness when infected by each pathogen genotype separately, weighted by the frequency of each of these in the pathogen population. Similar calculations provide the fitnesses of the various pathogen genotypes. As May & Anderson (1983) have pointed out, the interactive system produced is essentially one in which host fitnesses depend on relative gene frequencies in the pathogen population, and pathogen fitnesses depend on host gene frequencies. As a consequence, these and other models concerned with genetic changes in both host and pathogen populations (Mode, 1958, 1961) or in the pathogen population alone (Person *et al.*, 1976; Groth & Person, 1977) are able to maintain polymorphisms only if a fitness cost is associated with both virulence (g of Jayakar; k of Leonard) and resistance (f and c, respectively). Alternatively, the cost of resistance may be negligible if virulent pathogen genotypes cause less damage to susceptible hosts than those carrying resistance genes ($a > 0$; Leonard, 1977). These polymorphisms may be stable, cyclic or chaotic (see Levin, (1983) for a review of the more intricate mathematical models).

Agricultural multiline models

These are a special class of host–pathogen genetic model which have arisen in response to widespread concern that the use of mixtures or multilines for disease control in agriculture might, in fact, favour the rapid evolution of pathogen races with very broad ranges of virulence (so-called 'super-races'). Because only the pathogen population is treated as a dynamic variable (the size and genetic structure of the host population being fixed), in many respects these models are less satisfactory than those considered earlier. However, they do provide some insight into the critical interdependence that the number and distribution of resistance genes in the host population and the relative cost of unnecessary genes for virulence in the pathogen population have in determining the genetic structure of the latter population. By varying either of these parameters it is possible to determine whether or not an effective level of disease control can be achieved.

The simplest of such models (Groth, 1976; Groth & Person, 1977; Marshall & Pryor, 1978) are based on random mixtures of n diploid host genotypes, each homozygous for a different dominant gene conferring resistance to a specific haploid pathogen. These mixtures are envisaged as being grown over large areas and are reconstituted annually, so that their composition is stable over time. They act as the major selective force affecting evolution in the pathogen population. Furthermore, it is assumed: (1) that pathogen races with all possible combinations of virulence genes exist in the population; (2) that races carrying only necessary genes for virulence have equal fitness; (3) that each unnecessary virulence gene reduces fitness by a constant value k; and (4) that two or more such genes are either additive or multiplicative in their effects in reducing pathogen fitness. Under these conditions, the mean fitness of each pathogen race can be calculated and used to predict the composition of the population. Thus when the fitness cost associated with unnecessary virulence is additive, pathogen races virulent on all host genotypes will dominate when $k < 1/2(n-1)$. If, on the other hand, fitness costs are multiplicative, such a super-race will dominate when $k < 1/n$ (Marshall & Pryor, 1978).

Alternatively, these expressions can be reversed and used to determine the number of host genotypes which would have to be present in a mixture to prevent the dominance of a super-race. These relations are $n > (1+2k)/2k$ and $n > 1/k$ for the additive and multiplicative models, respectively. Clearly, in any particular host–pathogen combination the number of resistant genotypes needed to prevent the emergence of a super-race depends critically on accurate estimates of the cost of unnecessary virulence. Unfortunately this whole concept is fraught with great controversy and clouded by much conflicting evidence. If, however, we leave a thorough consideration of this problem until later (Chapter 8) and accept instead the generously wide-ranging estimates of k ($0.12 < k < 0.42$: Leonard, 1977) used by Marshall & Pryor (1978), we find that the number of host genotypes needed to prevent the dominance of a pathogen super-race is 6 ($k = 0.12$) or 3 ($k = 0.42$) in the additive fitness model and 9 or 3 in the multiplicative one.

In order to gain some insight into how non-agricultural host–pathogen systems work, a more realistic question to ask is just how many host genotypes are required to produce a community in which only 20% or even 40% of the host population is susceptible to any given pathogen race.

Again, unless the fitness costs associated with unnecessary genes for virulence are marked, the number of host genotypes required becomes large. Thus, for large k values ($k = 0.42$) both additive and multiplicative fitness models (Marshall & Pryor, 1978) needed only five or ten host genotypes to restrict individual pathogen races to either 40% or 20% of the total population. However, for lower values ($k = 0.12$) the two alternative models required, respectively, 13 and 20 host lines to limit individual races to even 40% of the host stand. A similar assessment has been made by Groth & Person (1977) who found that levels of diversity in general agreement with those of Marshall & Pryor (1978) were necessary to restrict their pathogen to 50% of the host population.

Further extensions of this basic model have examined the effect of the way in which resistance genes are incorporated into the host population on their ability to stabilize the pathogen population (Marshall & Pryor, 1979; Marshall & Burdon, 1981). Combinations of resistance genes are consistently superior or equal to their use singly only when these are incorporated into host genotypes in a non-overlapping fashion (i.e. AB, CD,...) and where unnecessary genes for virulence act additively to reduce pathogen fitness.

Several other models have been developed to examine host–pathogen genetic interactions in agricultural mixtures. These pay little attention to the host population, placing even greater emphasis on genetic changes in the pathogen population. They are considered in Chapter 8.

Combining numerical and genetic aspects of host–pathogen interactions

The host–pathogen models considered in this chapter are all frequency-dependent models in which the pathogen population is seen as affecting the relative abundance of host genotypes but not the absolute size or density of the population. Changes in the number of individuals of any particular genotype are expressed solely as changes in frequency.

The assumption of frequency dependence but density independence, while considerably reducing the complexity of interactions involved in any particular host–pathogen combination, fails to consider the effects that large-scale pathogen-induced changes in population size are likely to have on the genetic diversity of such populations. In dynamically active host–pathogen systems, long-term changes in the genetic constitution of a host population are likely to follow short-term fluctuations in the size and density of the population rather than to replace such changes. In fact, changes in the size of populations in response to pathogen pressure ('ecological feedback': Chilvers & Brittain, 1972) will invariably precede changes in the genetic structure of the population ('genetic feedback': Pimentel, 1961) unless the selective pressure exerted by the pathogen excludes individuals from the host population at exactly the same rate as resistant ones become established.

To date, no attempt has been made to extend frequency-dependent gene-for-gene models of the type developed by Jayakar (1970) or Leonard (1977) to include the effects of density dependence. However, May & Anderson (1983) have recently extended their analyses of the population dynamics of interacting populations of animal hosts and their parasites (Anderson & May, 1978, 1979, 1981, 1982; May & Anderson, 1978, 1979) to include a consideration of the effects of frequency- and density-dependent fitness functions on the genetic structure of such populations. Caution must be exercised in applying predictions from this animal host–parasite model to plant–pathogen co-evolutionary interactions. Many basic epidemiological features differ markedly between these two groups of associations; for example, May & Anderson's (1983) model was developed for a micro-parasite–host interaction in which individual hosts may recover and become immune to further parasite infection. However, it is interesting to note that, in line with the predictions of frequency-dependent models, polymorphism usually exists in the host population unless resistance and virulence have no associated fitness cost. While such polymorphisms are more liable to show cyclic or chaotic oscillations than are those with no parasite-induced density dependence (May & Anderson, 1983), additional complications (for example, cross-infection of hosts by the different pathogen genotypes) are likely to exert smoothing influences favouring stable polymorphisms over cyclic or chaotic ones.

As May & Anderson (1983) have pointed out, these theoretical models predict a range of different evolutionary pathways that can be followed by individual host–parasite associations. The particular direction taken depends upon details of the fitness cost of virulence in the parasite and resistance in the host. This is also true for plant–pathogen interactions. From the population biologist's point of view, pathways which lead to some kind of polymorphism in resistance and susceptibility in the host and virulence and avirulence in the pathogen are the most intriguing. These solutions apparently require the existence of costs associated with resistance and virulence. As a consequence, we turn here to a consideration of whether fitness costs are associated with resistance. A detailed consideration of the same question with respect to virulence is left until Chapter 8.

The cost of resistance

In contrast to the concept of a fitness cost being associated with virulence in pathogens, the possibility that a similar cost is also associated with resistance in hosts has received little attention. Arguing by analogy with studies on resistance to herbicides and heavy metals, one might expect marked effects of resistance genes on the general vigour and reproductive potential of plants. Studies of herbicide- or heavy metal-resistant and -susceptible forms of a range of plant species have generally found susceptible types to be competitively superior in the absence of the particular selective factor (herbicide: Conard & Radosevich, 1979; Weaver & Warwick, 1982; heavy metal: McNeilly, 1968; Hickey & McNeilly, 1975). Conard & Radosevich (1979) found coefficients of selection against S-triazine resistance of approximately 0.5 in both *Senecio vulgaris* and *Amaranthus retroflexus*, while McNeilly (1968) measured a value of 0.53 for a metal-tolerant form of *Agrostis tenuis* growing in competition with non-tolerant genotypes on uncontaminated soil. Similarly low fitness values have been associated with metal-tolerant forms of *Anthoxanthum odoratum*, *Plantago lanceolata* and *Rumex acetosa* when growing under competitive conditions on normal soil (Hickey & McNeilly, 1975).

If disease resistance genes uniformly carry comparable fitness costs, then it is reasonable to expect that the yield penalties associated with their use in agricultural crops would have been noted and investigated by plant breeders. This has not been the case. Rather, where yield costs have been detected and quantified, values of less than 10% have generally been recorded. In tobacco, statistically significant ($P < 0.05$) yield reductions of 5.6% and 6.5% were found when lines resistant to tobacco mosaic virus and *Fusarium* wilt, respectively, were compared with susceptible controls (Chaplin, 1970).

One problem which inevitably occurs in the interpretation of yield differences between resistant and susceptible lines is whether these differences are actually caused by resistance *per se* or whether they reflect the effect of other genes which have been carried along with resistance during selection. This problem is perhaps best illustrated by an experiment deliberately designed to compare the performance of resistant and susceptible lines of a host derived from a single natural population. In this study the reproductive output of 15 lines of *Avena fatua* susceptible to four races of *Puccinia coronata* (crown rust) was found to be greater, under some environmental conditions, than that of 15 resistant lines (Burdon & Muller, in press). Under other conditions, however, the resistant lines were more successful. Moreover, these differences in performance were apparently correlated with differences in the germination behaviour of the two sets of lines. These sorts of confounding associations are likely to be a constant problem in the interpretation of observations involving strongly inbreeding species like *A. fatua* where there is little opportunity for a random reassortment of characters in the population as a whole. Clearly it is difficult, if not impossible, to separate the fitness effects of individual characters like resistance from other associated features of the genome. We will return to this problem when considering temporal changes in the resistant structure of host populations.

Perhaps the best way to measure such fitness costs would be to monitor, over a number of generations, changes in phenotypic and genotypic frequencies in obligately outbreeding species that are segregating for resistance in disease-free environments. Such an approach should minimize, although not totally eliminate, linkage effects. Controlled experiments of this type have not been attempted but both van der Plank (1963) and Harlan (1976) have argued that resistance may be lost quite rapidly in plant species which are moved to areas in which particular diseases are not present. A much quoted example used to support this contention is that of maize which apparently became highly susceptible to attack by the pathogen *Puccinia polysora* (rust) during 300 years of isolation in Africa. Because many other plausible explanations exist for such observations, such anecdotal examples are, at the best, merely suggestive of a possible fitness cost associated with resistance.

While the preceding discussion has assumed that fitness costs associated with disease resistance would be expressed directly as differences in reproductive performance, this need not necessarily be the case. In fact, if these costs are apparent in some parameter other than yield (for example, survivorship), compensatory growth by other members of the population may result in total yields per unit area that are little different from those produced by fully susceptible controls. This would decrease the likelihood of their detection by plant breeders.

A more convincing test of both the basic assumptions and the predictions arising from the theoretical arguments developed in gene-for-gene population models lies in a consideration of empirical data derived from investigations of real host–pathogen interactions in non-agricultural communities.

Empirical evidence

Variation in resistance within populations

Pitfalls in assessing disease resistance

The apparent extent of variation for disease resistance in wild plant populations is greatly influenced by the degree of discrimination which can reliably be achieved between different infection type responses. This, in turn, varies greatly between different host–pathogen combinations. In some combinations insufficient knowledge is available concerning either host or pathogen, and infection type responses can safely be assigned to two or three categories only. In other cases, particularly those involving graminicolous hosts and their pathogens, it has been possible to use infection type assessment scales derived from agriculture. Thus studies of the interaction between *Erysiphe graminis* (powdery mildew) and *Hordeum spontaneum* (Fischbeck *et al.* cited by Wahl *et al.*, 1978; Moseman, Nevo & Zohary, 1983) or *Triticum dicoccoides* (Moseman *et al.*, 1984) used a standard 0–9 semi-continuous scale to classify the infection type responses of individual members of host populations. These categories ranged from immunity (0) to complete susceptibility (9). In one population of *T. dicoccoides* growing at Tabigha in Israel, 47% of individuals were resistant to a mixture of two races of *E. graminis* and 34% were susceptible, while the remaining 19% showed intermediate infection type responses. These were characterized by lesions of varying size and leaf necrosis or chlorosis of varying intensity (Moseman *et al.*, 1984). Similar ranges of variability in infection type response were found in individual populations of *Hordeum spontaneum*. In one of these, Fischbeck and his colleagues also assessed the occurrence of a 'slow mildewing' presumably race non-specific type of resistance. This was apparent as differences in infection severity on adult plants. Infection severity values on different plants ranged from a trace to 50% of the leaf surface being covered with fungal colonies. Individuals which showed resistant or intermediate seedling infection type responses all suffered severities of less than 15% cover. However, individuals with susceptible seedling responses showed a wide diversity of slow mildewing resistance (5–50% severity ratings) (Fischbeck *et al.* in Wahl *et al.*, 1978).

Striking variation in resistance to disease has also been recorded in populations of a range of other species (see Table 7.3), including forest trees like *Pinus taeda*. Variability in the response of a population of this species to the pathogen *Cronartium fusiforme* (fusiform rust) has been assessed by examining the occurrence and severity of disease on half-sib families derived from randomly chosen parent trees growing in a natural

Table 7.3 *Representative examples of the range and mean level of resistance to fungal pathogens found in a variety of host–pathogen combinations (after Burdon, 1985)*

Host–pathogen combination		Number of populations	Mean sample size per population	Range of resistance[a] (%)	Mean level of resistance (%)	Number of races[b]	Comments[c]	Reference
Avena sp./ *Puccinia coronata*		31	20	0.0–?	12.4	2	SR	Dinoor, 1970
	(i)[d]	11	44	0.0–95.0	27.3	4	SR	Burdon *et al.*, 1983
	(ii)[d]	10	46	0.0–5.0	1.0	4	SR	Burdon *et al.*, 1983
Helianthus annuus/ *Puccinia helianthi*		142	*c*20	0.0–*c*80.0	12.8	1		Zimmer & Rehder, 1976
Helianthus petiolaris/ *Puccinia helianthi*		57	*c*20	0.0–100.0	85.4	1		Zimmer & Rehder, 1976
Hordeum spontaneum/ *Erysiphe graminis*		15	17	0.0–100.0	50.8	composite	SR	Moseman *et al.*, 1983
		77	35	0.0–45.0	?	composite	AR	Fischbeck *et al.*, 1976
Pinus taeda/ *Cronartium fusiforme*		3	11	12.1–21.2	16.1	composite		Barber, 1966
Triticum dicoccoides/ *Erysiphe graminis*		10	23	16.7–100.0	55.3	composite	SR	Moseman *et al.*, 1984
Phlox pilosa detonsa/ *Erysiphe cichoracearum*		10[e]	19	57.2–96.3	77.1	1		Jarosz, 1984
Phlox drummondi mcallisteri/ *Erysiphe cichoracearum*		4[e]	26	43.4–74.6	62.3	1		Jarosz, 1984
Phlox pulcherrima/ *Erysiphe cichoracearum*		5[e]	13	63.4–83.3	74.0	1		Jarosz, 1984

[a]Resistance measured in a variety of ways; see original references for explicit details. [b]Number of pathogen races inoculated individually. [c]SR = seedling resistance; AR = adult plant resistance. [d](i) and (ii) = environments favourable and unfavourable for the pathogen, respectively (see text). [e]Including only those populations with a sample size $\geqslant 10$.

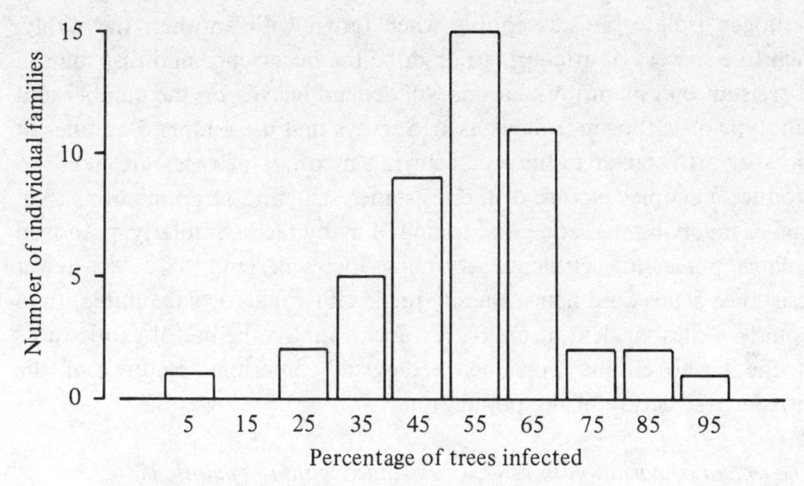

Figure 7.1. The response of a population of *Pinus taeda* to infection by the pathogen *Cronartium fusiforme* (fusiform rust). The distribution shows the mean percentage of trees infected per family (approximately ten plants/family). (Redrawn from Kinloch & Stonecypher, 1969.)

stand (Kinloch & Stonecypher, 1969). After 6 years' exposure to natural infection the severity of disease on different family lines showed a normal distribution (Figure 7.1). One family was highly resistant (< 5% of progeny were infected); another was highly susceptible (> 95% were infected), and 87% of the remainder possessed intermediate levels of resistance (30–70% infected). As this resistance was highly heritable, the original parental trees clearly varied greatly in their response to this pathogen. Moreover, the resistance observed almost certainly reflected the interaction of a wide range of mechanisms operating at different morphological and biochemical levels in *P. taeda* to reduce the infection and severity of *C. fusiforme* (Bingham, Hoff & McDonald, 1971; Kinloch, 1982).

Investigations of this type indicate that considerable variability *may* occur between individual members of a population in their resistance to particular diseases. At the same time, however, studies which use only one or two isolates or mixtures of pathogen cultures can provide an incomplete picture of the complexity of the resistance structure of many plant populations. The basic interdependence of the phenotypic expression of resistance or susceptibility in the host with avirulence or virulence in the pathogen ensures that resistance genes are only expressed phenotypically when challenged by avirulent pathogen isolates (see Chapter 4). As a result, any individual host may be classified as resistant when challenged with one pathogen isolate but susceptible when tested with another. Inevitably, then, the success of attempts to describe the occurrence and distribution of resistance in plant populations will depend heavily on the number and genotype of pathogen isolates used. Surveys that use a limited number of races or artificial or naturally occurring mixtures of races are likely to produce a simpler picture of the resistance structure of populations than studies involving the sequential testing of many races. Similarly, pathogen isolates possessing virulence at many loci will tend to detect fewer resistance genes (and hence classify more individuals as susceptible) than would isolates virulent at only a few loci. Finally, the use of exotic races of the pathogen may produce ecologically spurious pictures of the resistance structure of the population.

The extent of variation in disease resistance within populations

Perhaps the best solution to the problems detailed above is to screen populations for resistance with a number of the commoner races occurring naturally in the study area. This approach has been used successfully to investigate the response of a number of Australian populations of *Avena barbata* and *A. fatua* to four races of *Puccinia coronata* (crown rust) (Burdon, Oates & Marshall, 1983). In this study the infection type responses of individual members of each population were divided into seven categories ranging from complete resistance (1) to complete susceptibility (7). In a typical population the infection type responses of seedlings derived from 50 *A. fatua* plants produced resistance profiles which were similar for all four pathogen races (Figure 7.2). A few host individuals were either highly resistant or highly susceptible to the particular pathogen race involved, while the majority showed intermediate levels of resistance. While the infection type responses of most hosts in this population were the same for all races a few showed differential reactions. In other populations the proportion of individuals showing differential reactions was greater.

The complexity that these patterns of interaction between host and pathogen populations may attain is illustrated by a study of resistance to *Phakopsora pachyrhizi* (rust) in a natural population of the Australian native legume *Glycine canescens*. The phenotypic response of 22 plants to nine different races of the pathogen was determined by recording the infection type responses of 10–20 seedlings per host individual–pathogen race combination (Figure 7.3; Burdon, unpublished data). As found previously with the *Avena* studies, individual host plants showed considerable diversity in their response. Half of the individuals were resistant to

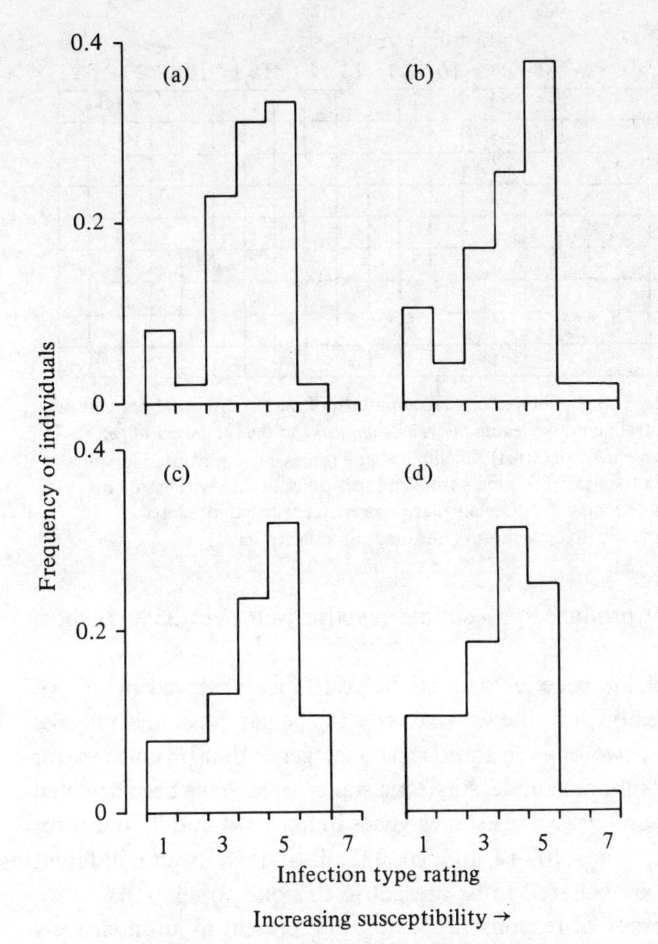

Figure 7.2 The frequency distribution of the infection type response of an *Avena fatua* population (growing at Inverell, New South Wales, Australia) to four different races of the pathogen *Puccinia coronata* (crown rust). (a) Race 226; (b) Race 227; (c) Race 237; (d) Race 277. Infection type rating of 1 equals full resistance; 7 = full susceptibility. (From Burdon *et al.*, 1983 and unpublished data.)

all races of the pathogen, while others possessed resistance effective against 11–88% of the races used. No individuals were susceptible to all races nor were any pathogen races virulent on all host lines. These results also provide a clear example of the different levels of resistance that different pathogen isolates can detect within a single host population. Race 1 of *Phakopsora pachyrhizi* is relatively avirulent and produced a susceptible reaction on only 4.5% of the population. At the other extreme, virulent

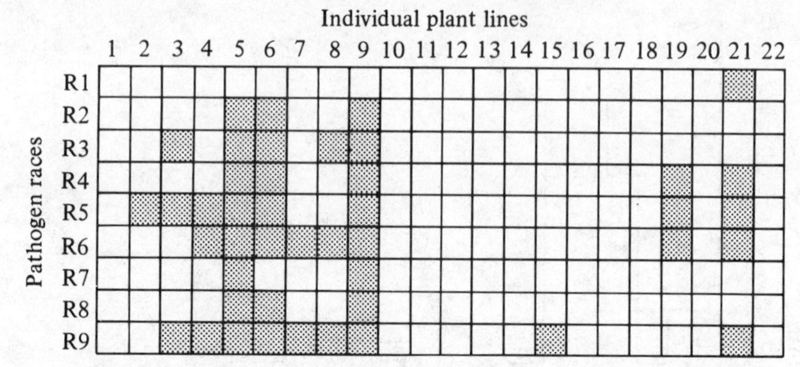

Figure 7.3 The phenotypic resistance pattern of part of a natural population of the Australian native legume *Glycine canescens* to the fungal pathogen *Phakopsora pachyrhizi* (rust). Each plant line represents a separate individual collected in the field. The nine pathogen races differ in their virulence on a standard differential set. (Open squares = resistance reaction; closed squares = susceptible reaction.) (Burdon, unpublished data.)

races 5, 6 and 9 produced susceptible reactions on over 35% of host individuals.

An analysis of the genetic basis of the resistance observed in this *G. canescens* population has shown that any particular host line may be protected by one, two or even three resistance genes that produce major phenotypic effects. For example, single resistance genes have been detected in host lines 2, 5 and 9; two resistance genes in lines 3, 4 and 7; and three different genes in lines 10, 14 and 20. In all, at least twelve different resistance genes are believed to be present in this one population.

A similar diversity of response appears to be present in an intensively studied *Avena* population from Israel (Dinoor, 1977). Seedlings derived from 109 plants growing in this stand were classified into one of a number of combinations of resistance and susceptibility on the basis of their infection type response to six common races of *Puccinia coronata*. While 24% of the population was found to be susceptible to all six races, the remaining 76% was distributed among 17 different combinations of resistance and susceptibility. Some of these combinations occurred at high frequency while others were represented by single individuals only. For example, 25 individuals were resistant to races 264, 276 and 277 alone, seven others were resistant to these three races and also race 202, but only one plant was resistant to all six races tested.

Overall, such populations present their pathogens with complex two-dimensional mosaics of resistance and susceptibility. Not only can individual host plants react differently to different pathogen races, but the total fraction of the population which is susceptible also varies with pathogen phenotype (Burdon, 1985). Moreover, these features may also vary with time as the ontogenetic age of individual plants changes. As we will see, however, not all plant populations show a diversity of response to individual pathogens. Many populations are totally susceptible while in others all members appear to be fully resistant.

It must always be remembered that resistance mechanisms other than those characterized by differences in infection type responses in seedlings or adult plants undoubtedly also exist in natural plant populations. As Dinoor & Eshed (1984) have pointed out, even in populations in which race specific forms of resistance are uncommon, relatively little pathogen-induced damage may occur as a result of the effects of a variety of race non-specific resistance mechanisms that are expressed throughout the life of a plant (see Chapter 4). However, because of the difficulty of assessing large numbers of host lines individually for such epidemiologically based characters as 'slow' rusting or mildewing or tolerance, virtually nothing is known about the frequency of these types of resistance (with the notable exception of the earlier mentioned work of Fischbeck and his colleagues) or of their relative contribution to the overall response of the population to pathogen-induced selection pressures. Similarly, it is not known whether early maturity in wild plants has been favoured as a direct consequence of its disease avoidance potential or whether this was a fortuitous, but incidental, benefit of selection in response to an unpredictable physical environment.

The significance of variation in the frequency distribution of resistance

Does the frequency distribution of infection types in a population provide us with clues concerning the nature and extent of the selection pressure placed on individual plant populations by fungal pathogens? In a study of the response of a population of *Trifolium repens* to attack by the two leaf pathogens *Cymadothea trifolii* (black blotch) and *Pseudopeziza trifolii* (leaf spot), the frequency distribution of resistance to the first of these pathogens was normally distributed; that to the second pathogen was strongly skewed towards greater resistance (Burdon, 1980). It was tentatively suggested that these differences reflected the relative importance of the two pathogens as selective forces operating on the host population in the recent past. *P. trifolii* was postulated to have had a more severe effect on the host population than *C. trifolii*. In an opposing view, Dinoor & Eshed (1984) have cited various examples involving rusts and mildews of

Avena sterilis and *Hordeum spontaneum*. In these species, apparently normal distributions have been associated with high disease resistance in the field while distributions skewed towards resistance have been associated with the use of exotic pathogen cultures.

Overall there appears to be some merit in both of these arguments. However, a number of factors are likely to affect the frequency distribution of infection types in a host population. These include: (1) the history of pathogen attack – its frequency, severity and the racial composition of the pathogen populations involved; (2) the genetic basis of the resistance present in the host – whether this is under simple or polygenic control; (3) the cost of resistance; and (4) the nature of the breeding system of the host species in question. The last of these factors encompasses one of the most obvious differences between *Trifolium repens* and *Avena sterilis* and *Hordeum spontaneum*. *T. repens* is a self-incompatible species in which disease resistance is likely to be selected independently of other characters in the genome. As a result, differences in the frequency distribution of resistance may well reflect differences in selective pressures exerted by different races or species of pathogen. On the other hand, in strongly inbreeding species like *A. sterilis* and *H. spontaneum*, co-adapted gene complexes are a common feature (Brown, Feldman & Nevo, 1980) and strong selection for any one particular aspect of the complex may cause at least short-term non-adaptive changes in other associated characters. This would potentially permit the development of skewed frequency distributions even in situations where disease selection pressures are not particularly high. These sorts of complications must be borne in mind when assessing the significance of resistance patterns in plant populations.

Variation in resistance between populations

Different populations of a wide range of plant species often exhibit marked differences in resistance when tested under uniform conditions (see Table 7.3). While the level of resistance possessed by adjacent populations may vary considerably, the distribution of resistance is frequently not random. Populations with similar levels of resistance (high or low) are often associated with one another in particular geographical regions. In studies where this has been observed, differences have generally been linked with variations in the physical environment and thus, by inference, with changes in the prevalence of particular pathogens. The distribution of resistance in Israeli populations of *Avena sterilis* to *Puccinia coronata* (crown rust) (Dinoor, 1970; Wahl, 1970), of *Hordeum spontaneum* to *Erysiphe graminis hordei* (powdery mildew) (Fischbeck *et al*., 1976;

Moseman *et al.*, 1983) and of *Triticum dicoccoides* to *E. graminis tritici* (Moseman *et al.*, 1984) all show marked regional variations that have generally been associated with differences in moisture regimes and patterns of host maturity. In the drier southern regions, environmental conditions are often marginal for the sustained growth and development of either *E. graminis* or *P. coronata* and host populations in these areas tend to be composed mainly of susceptible individuals. By contrast, in many parts of northern Israel conditions favourable for the development of these pathogens coincide with conditions favourable for growth of host populations (Wahl, 1970). These populations are typically characterized by considerably higher frequencies of resistant individuals. An exception to this concentration of resistance in northerly Israeli populations of wild cereals is found in the distribution of resistance in *Avena barbata* populations to *P. coronata*. Although regional differences in the frequency of resistance were again detected, resistance tended to occur more frequently in the southern part of the country (Dinoor, 1970). Why this species should behave so differently from other wild cereals (particularly, *A. sterilis*) is not clear, although distinct differences between the two *Avena* species in their growth patterns may be at least partially responsible.

The effect that interactions between hosts, pathogens and their environment have on the distribution of resistance has been examined in more detail in an Australian study of wild oat (*Avena* spp.) populations. Again, resistance was found more frequently in some regions than others; the mean level of resistance of populations of *A. barbata* and *A. fatua* to four races of *P. coronata* and the overall diversity of the resistance profiles of individual populations growing in New South Wales was greater in the north than in the south (Burdon *et al.*, 1983). The extent of these differences is to be seen in the resistance profiles of representative *A. fatua* populations illustrated in Figure 7.4. Like the Israeli studies, these regional differences in the frequency of resistance could also be associated with differences in rainfall and temperature patterns and other climatic features. However, a detailed study of the pathogen population uncovered a further level of complexity. The racial diversity and the mean range of virulence of pathogen isolates collected in northern areas was also greater than those from the south (Oates, Burdon & Brouwer, 1983).

Clearly a complex range of interactions occurring between hosts, pathogens and their environment may be hidden behind simple observations of regional variation in the frequency of resistant individuals in plant populations. In the present case, the selective pressure exerted by *P. coronata* on its hosts in southern New South Wales appears to be limited

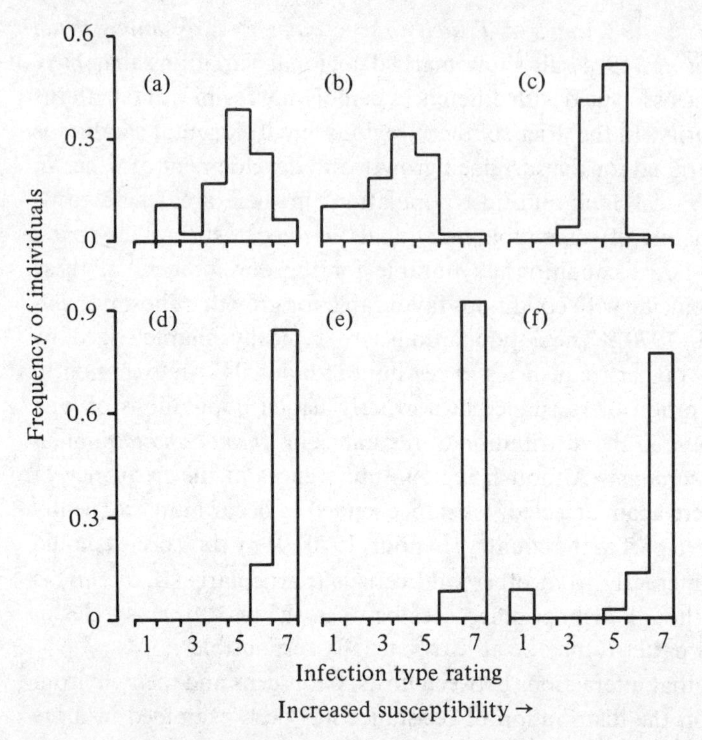

Figure 7.4 Frequency distributions of the infection type ratings of six *Avena fatua* populations to *Puccinia coronata* (crown rust) race 227. Infection type rating of 1 equals full resistance; 7 = full susceptibility. Populations are from northern New South Wales: (a) Glen Innes; (b) Inverell; (c) Warialda; or from southern New South Wales: (d) Murrumburrah; (e) Bowning; (f) Canberra. (Redrawn from Burdon *et al.*, 1983.)

by environmental conditions that are generally unfavourable for prolonged, simultaneous development of the two populations. In the northern region, on the other hand, conditions are favourable for longer and the two populations appear to be evolving in response to one another.

Associations between levels of resistance and environmental conditions have been taken beyond the highly generalized associations found in the studies described above by a study of 112 populations of ten species or sub-species of *Phlox* growing wild in the central and southern United States (Jarosz, 1984). Statistically significant, but modest, correlations ($r = 0.19$) occurred between the mean level of resistance shown by these populations to one isolate of *Erysiphe cichoracearum* (powdery mildew) and latitude or mean daily temperature. Resistance was more commonly encountered in more northerly and cooler environments. When 65 of these

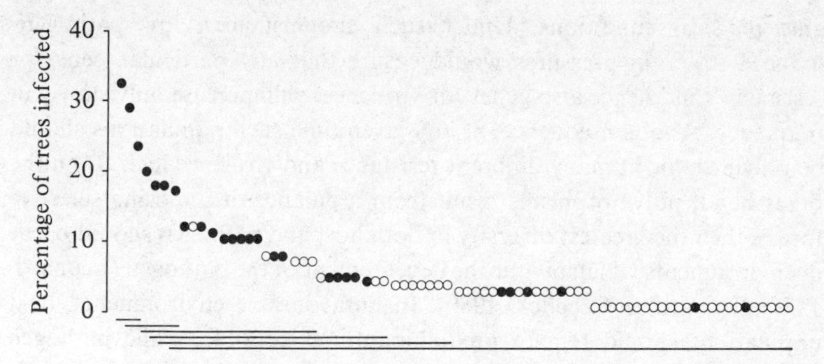

Figure 7.5 *Cronartium comptoniae* (sweet fern rust) infection of *Pinus contorta* provenances from within (○) and beyond (●) the range of the alternate host, *Myrica gale*. Provenances underscored by the same line are not significantly different ($P = 0.05$). (Redrawn from Hunt & Van Sickle, 1984.)

populations were subjected to a detailed environmental assessment, Jarosz found a much stronger correlation ($r = 0.34$) with a habitat rating based on the amount of tree cover and the direction of exposure of the site. This rating scale reflected a gradient of decreasing aridity – falling maximum temperature and increasing relative humidities, which provide conditions increasingly suitable for development of the pathogen.

Changes in physical environmental conditions need not be the only features that affect the relative severity of disease and hence the frequency of resistance in different populations. In a survey of the resistance of 77 provenances of *Pinus contorta* to the rust *Cronartium comptoniae* (sweet fern rust), Hunt & Van Sickle (1984) found that the occurrence of resistance was strongly correlated with the geographical distribution of *Myrica gale*, the alternate host of this heteroecious fungus (Figure 7.5). The frequency of trees that became infected during 5 years' field exposure to *C. comptoniae* was significantly lower in provenances drawn from within the geographic range of *Myrica gale* than in those from beyond its range (average of 2% *versus* 10%; $P < 0.01$). A similar association was found between the distribution of resistance to *C. comptoniae* in *Pinus banksiana* and the occurrence of another alternate host, *Comptonia peregrina*.

These sorts of results provide some empirical verification of one of the theoretical predictions of the gene-for-gene population models considered earlier. If no fitness costs are associated with the possession of either resistance or virulence, polymorphisms in both host and pathogen will be transient. The greatest diversity of genotypes should then occur in environments where pathogen activity is restricted by generally unfavourable physical conditions. Under such circumstances, low pathogen-induced selection pressures would ensure that any particular gene for resistance (and hence also genes for virulence) will increase only slowly in frequency. As a consequence, at any given time such populations should be polymorphic at many different resistance and virulence loci. If, on the other hand, polymorphisms result from a balance of opposing selective forces, then the greatest diversity in both host and pathogen should occur in environments which favour the development of the pathogen (Leonard, 1977; Leonard & Czochor, 1980). In unfavourable environments, host populations should remain predominantly susceptible, while pathogen populations should remain simple. Theoretical aspects of this argument have been extensively developed by Leonard (1984) in an extension of his earlier models. Evidence of the type collected in Israel and Australia provides support for a system of balanced polymorphisms in host and pathogen populations.

Variations in resistance on a local scale

Even in regions generally suited to the extensive development of particular pathogens, considerable variation may occur between adjacent populations in the frequency of resistant individuals. The survey of Moseman and his associates (1983) is sufficiently detailed to illustrate this point. Over 90% of individuals growing in a *Hordeum spontaneum* population at Rosh Pinna in northern Israel were resistant to two composite cultures of *Erysiphe graminis hordei* (powdery mildew). Populations growing 10 km north-west at Gadot or the same distance south at Tabigha were both more susceptible. The actual frequency of resistant, susceptible or intermediate types in these populations depended on the particular culture of *E. graminis hordei.* In the most extreme case, over 75% of individuals were susceptible.

Uniformly resistant or susceptible plant populations may also occur as a result of genetic bottleneck effects during their establishment. This is most likely in highly inbreeding or agamospermous species growing in marginal or highly ephemeral habitats. In such cases, the establishment of a single seedling may lead to the development of a highly uniform population, all members of which have the same response to particular pathogen races as the founding individual. In the short term, because there is no variation for selection to act upon, the entire population will be either resistant or susceptible, regardless of the level of pathogen activity. This establishment effect appears to be the main factor involved in the observed differences between fugitive populations of *Lactuca serriola* in England. While considerable variation was observed between populations of this species, within each population all individuals responded uniformly to the pathogen *Bremia lactucae* (downy mildew) (I. R. Crute, personal communication 1982).

The extent to which local differentiation may occur in the resistance structure of host populations has not been investigated. However, in theory at least, a lower limit to the scale of local selectively relevant differentiation is determined by the rapidity with which selection pressures exerted by pathogens alter across environmental gradients and the degree to which these are offset by gene flow between adjacent populations. Thus, for example, in an interaction involving a pathogen particularly favoured by moist, humid conditions, host populations growing on an exposed, dry hillside would be likely to suffer far less selection pressure from disease than would populations of the same species growing in a damper, humid valley bottom a short distance away (see Chapter 5 for a full discussion of the effect of the environment on the occurrence and severity of disease). In such a scenario, hillside populations might be expected to remain susceptible while valley populations may rapidly become more resistant. Although gene flow between adjacent populations will tend to counteract such local differentiation many examples exist of abrupt changes in population structure despite marked gene flow (e.g. heavy metal-tolerant/sensitive populations of *Agrostis tenuis* (McNeilly, 1968) and *Anthoxanthum odoratum* (Antonovics & Bradshaw, 1970)).

To date only a few studies have shown differentiation in the disease resistance structure of plant populations at the same local scale as found in studies of heavy metal tolerance. One of these involves differences between populations of *Anthoxanthum odoratum* occurring on differently treated plots of the Park Grass Experiment at Rothamsted, England (Snaydon & Davies, 1972). Samples of this species taken from plots subject to a range of management practices showed considerable variation in many morphological and physiological characters as well as resistance to *Erysiphe graminis* (powdery mildew) and *Puccinia poae-nemoralis* (rust). Resistance to *E. graminis* was closely correlated with total soil nitrogen levels, disease severity being lowest and hence resistance highest in populations derived from plots with high soil nitrogen. Similarly, populations of *A. odoratum* taken from plots with tall vegetation (and thus more humid micro-climates) were more resistant to *P. poae-nemoralis* than were those from plots with short vegetation. In both cases, the most resistant populations were found growing on plots where environmental conditions particularly favoured the growth and development of the respective pathogen (i.e. high soil nitrogen, high humidity).

These differences in resistance were detected between populations 30 m apart. However, subsequent studies of characters like the time to anthesis, panicle height and yield showed significant, genetically based differences between populations separated by only 10 cm at the interface between adjacent plots (Snaydon & Davies, 1976). Unfortunately, this study did not consider the disease response of these individuals. However, the marked change in selection pressures occurring at these boundary zones may well be sufficient to produce similar micro-scale differences in the resistance structure of contiguous populations.

Another study documenting local scale differentiation in disease resistance involves an interaction between six sub-populations of the inbreeding annual legume *Amphicarpaea bracteata* and the pathogen *Synchytrium decipiens*. The host sub-populations were 30 m apart. When challenged with one isolate of the pathogen, five sub-populations displayed a high level of susceptibility (> 80%). The sixth was totally resistant (Parker, 1985). This difference is again interesting, although it may simply reflect the tight, inbred descent of individual sub-populations from individual founder genotypes. The selective relevance of the difference could perhaps be tested by experimental manipulation of the host–pathogen interaction. This approach has been used by Jarosz (1984) to demonstrate the selective basis of a strong association between increasing shadiness of a habitat and increasing resistance in *Phlox* spp. to *Erysiphe cichoracearum* (powdery mildew). Over a 3-year period, experimental populations of susceptible individuals planted into a shady habitat were consistently more heavily diseased than similar populations established in an exposed site 25 m away. These differences in disease severity were translated into differences in the survival and reproductive performance of individual plants (Jarosz & Levy, personal communication 1983).

While differences in the level of resistance occurring in adjacent plant populations are of considerable interest, we are still left with a question – over what spatial scale are true co-evolutionary interactions involving reciprocal genetic changes in *both* host and pathogen likely to occur? The currently available models of co-evolutionary interactions between host plants and their pathogens are not very helpful in this regard. These models assume that individual host populations are uniformly distributed and large enough to maintain the pathogen population indefinitely. However, they give no indication as to what this size might be. In reality, most environments are not homogeneous and the entire host population within an area is likely to be dispersed among many small patches (sub-populations) that are spatially separated to varying degrees. As was pointed out in Chapters 3 and 6 this will have marked effects on both the occurrence of a pathogen in any given patch and the size of the pathogen population. In addition, such sub-structuring of the population has important implications for the spatial scale at which co-evolutionary interactions may be expected to occur.

If host patches are small and short-lived, or periodically do not support the pathogen, then the pathogen population as a whole must survive by way of a fugitive strategy. It must continually invade new patches only to become locally extinct as conditions change or the hosts die. In such circumstances, individual highly localized associations of host and pathogen are unlikely to exist for long enough to allow reciprocal genetic changes to occur. Furthermore, the relatively high dispersal capacity of many pathogens is likely to ensure that the virulence spectrum of at least some of the isolates present in any given host patch will reflect selective pressures exerted by hosts in other patches (see Chapter 8). These two factors will tend to ensure that reciprocal, co-evolutionary interactions between hosts and pathogens will occur over areas that are much larger than the individual patch.

Variation in resistance with time

The importance of aspects of the physical environment in determining the long-term outcome of competitive interactions between mixtures of different plant genotypes has been ably demonstrated on a number of occasions (e.g. Harlan & Martini, 1938; Suneson & Wiebe, 1942). In contrast, very little empirical evidence is available concerning the long-term effects of disease on the frequency of resistant and susceptible phenotypes. In fact, such information is restricted to two studies: one which monitored the compositional stability of an oat multiline; and one which recorded changes in populations of the weedy species *Chondrilla juncea* during the course of a biological control programme.

The first of these studies was a controlled field experiment in which the possible effects of differential competitive abilities between the component lines of the mixture, and their response to year-to-year variations in climatic conditions, were separated from potential selective changes induced by differential pathogen pressure (Murphy *et al.*, 1982). Five near-isogenic lines of *Avena sativa* were mixed together in equal proportions to produce an initial population which was then divided in two and grown for four consecutive years. At the end of each generation a random sample of seed was collected and used to produce the subsequent generation. Each year, half the experiment was inoculated with a mixture of five races of the rust pathogen *Puccinia coronata* (crown rust), while the other half was protected from this pathogen by repeated fungicide treatment. Over the four generations of the experiment, changes in the frequency of individual lines in the two environments were generally similar. None of the differences between treatments were statistically significant. However, in the rust-free plots, despite considerable year-to-year variation in the relative success of different genotypes, there was a distinct suggestion that some would eventually dominate the mixture while others would be eliminated or at least reduced to low frequencies. From an ecological point of view, it is a great pity that changes in the composition of this mixture were not followed for longer.

The apparent resilience of the structure of this mixture to disease-induced selection pressures is intriguing but possibly reflects the experimental design. In the original mixture, the resistance possessed by each genotype overlapped so that any particular line was susceptible to only one of the five pathogen races (Murphy *et al.*, 1982). From the perspective of each host and pathogen genotype, the initial population was effectively 80% resistant. Even after four generations, the highest frequency of an individual oat genotype in either mixture was still only 38% (in the rust-free plot). At this frequency the disease-reducing effects of most mixtures are still highly effective and are likely to reduce substantially the selective pressure exerted by specific pathogen races (see Chapter 3). Moreover, as all host genotypes were susceptible to at least part of the pathogen population that occurred each year, any marked pathogen-induced increase or decrease in the frequency of a particular host genotype is likely to be followed by a similar equilibrating change in the frequency of the appropriate pathogen race.

Results of attempts to control the weedy composite species *Chondrilla juncea* in Australia using a deliberately introduced strain of the rust pathogen *Puccinia chondrillina* give some indication of the intensity of effect that pathogen-induced selective forces may have in a field environment. *Chondrilla juncea* is an obligately agamospermous species, numerous genotypes of which occur within its native range in Mediterranean Europe and Asia Minor. In Australia, three distinct forms have been introduced and by the late 1960s one of these, the narrow-leaved form, was widespread throughout much of south-eastern Australia. The other two (intermediate and broad-leaved) forms had more restricted distributions (see Cullen &

Figure 7.6 Natural heavy infection of the rust *Puccinia chondrillina* on a plant of the narrow-leaved form of *Chondrilla juncea* growing in the field in New South Wales, Australia.

Groves, 1977; Groves & Cullen, 1981). At this time, a survey of seven populations growing in New South Wales found most of these to be dominated by the narrow-leaved form. By 1980, however, 9 years after the successful introduction of a strain of *Puccinia chondrillina* virulent on only

Table 7.4 *Percentage of plants of three genotypes of* Chondrilla juncea *in local roadside populations in southern New South Wales, Australia. 1968 – before the release of a rust pathogen* (Puccinia chondrillina) *capable of attacking the narrow-leaved genotype*; *1980 – nine years after the release of this pathogen. N, I and B are the narrow-, intermediate and broad-leaved genotypes, respectively.* (*From Burdon* et al., *1981*)

	C. juncea genotypes %					
	1968			1980		
Population	N	I	B	N	I	B
1	100	0	0	88	0	12
2	100	0	0	99	1	0
3	98	0	2	5	90	5
4	67	30	3	3	78	19
5	62	24	14	3	80	17
6	51	10	39	8	10	82
7	21	1	78	11	13	76

the narrow-leaved form (Figure 7.6), the frequencies of all three forms of *Chondrilla juncea* in these populations had changed markedly (Table 7.4; Burdon, Groves & Cullen, 1981). At sites previously occupied by the narrow-leaved form and either one or both of the other forms, an increase in frequency of the latter forms reflected a decline in the frequency of narrow-leaved individuals. In the most extreme case (population 3), the frequency of the susceptible, narrow-leaved form fell from 98% to 5% while that of the intermediate form rose from 0% to 90%. These changes in the structure of populations inhabiting particular sites were also paralleled by changes in the distribution and abundance of all three forms on a wider geographical scale (Burdon *et al.*, 1981). As a result of the effects of the pathogen, the abundance of the widespread narrow-leaved form declined substantially. In contrast, the other two resistant forms spread well beyond the boundaries of their original distribution and are becoming major weeds in their own right.

Changes in resistance and the breeding systems of host populations

A major problem in the interpretation of long-term changes in the frequency of resistant individuals in plant populations is to determine whether such patterns truly reflect changes in selection for resistance *per se* or selection for associated features of the genome. In studies of self-incompatible species this is unlikely to present serious difficulties. Each

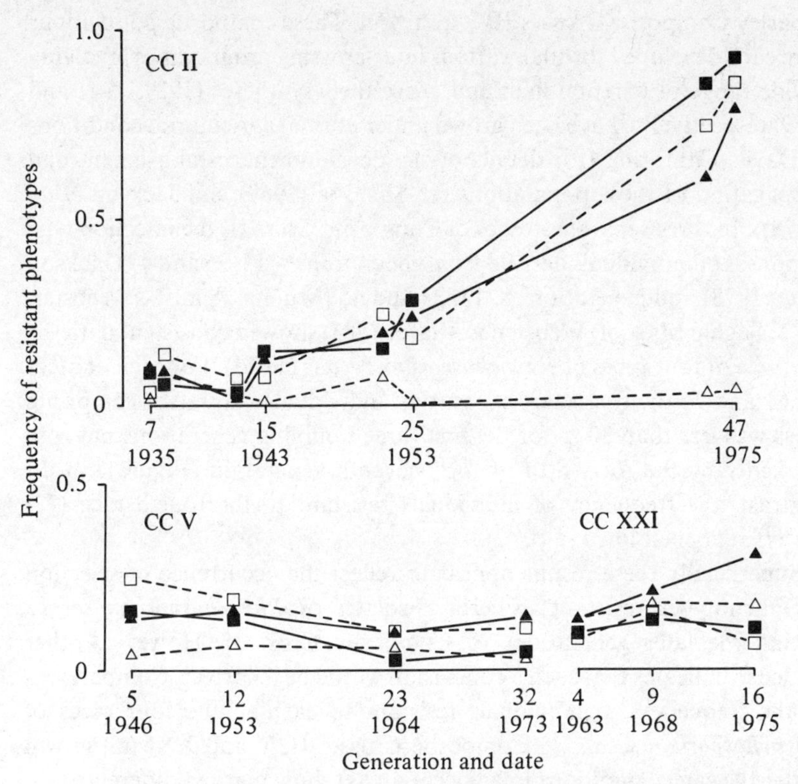

Figure 7.7 The frequency of individuals in various generations of Composite Crosses II, V and XXI resistant to four different races of the pathogen *Rhynchosporium secalis* (barley scald). Race 40, □; race 61, ▲; race 72, △; race 74, ■. (Data from Jackson *et al.*, 1978; Muona *et al.*, 1982; Saghai Maroof *et al.*, 1983.)

individual in populations of such species is likely to be genetically unique and differences like those observed between populations of *Anthoxanthum odoratum* can generally be assigned with confidence to differences in the intensity of selection for resistance (Snaydon & Davies, 1972). In strongly inbreeding species, however, this problem is of considerable importance. Populations of these species are generally divided into one or more families, members of which have largely the same multilocus genotype (Allard *et al.*, 1972; Clegg, Allard & Kahler, 1972; Brown *et al.*, 1980). In such populations, changes in the frequency of resistance alleles result from selection acting on the genotype as a whole rather than on individual resistance genes. As a result, separating the effects of selection for resistance *per se* from that for entire genotypes may be extremely difficult.

This problem is particularly illustrated by studies of pathogen resistance in barley Composite Crosses II, V and XXI. These composite populations were each developed through various inter-crossing programmes involving a wide range of parental lines and, since their syntheses (1929, 1941 and 1959, respectively), have been grown under normal agricultural conditions at Davis, California (for details of the development, establishment and propagation of these populations see Suneson (1969) and Jackson *et al.* (1978)). In three separate studies of Composite Cross II, the infection type responses of individuals derived from generations 7, 15, 25 and 47 (Jackson *et al.*, 1978) and generations 8, 13, 23 and 45 (Muona, Allard & Webster, 1982; Saghai Maroof, Webster & Allard, 1983) showed consistent patterns for three of four races of *Rhynchosporium secalis* (scald). For each of races 40, 61 and 74 the frequency of resistant individuals within the composite cross was less than 30% for the first three sampling generations but rose markedly by the forty-fifth or forty-seventh generation (Figure 7.7). In contrast, the frequency of individuals resistant to the fourth race (72) always remained low ($< 10\%$).

Superficially these results appear to reflect the occurrence of selection for phenotypes resistant to certain isolates of *Rhynchosporium secalis* during the later generations of Composite Cross II. However, other evidence indicates that such a conclusion would be incorrect. Comparisons of the frequency of individuals resistant to each of the four races of *Rhynchosporium secalis* in Composite Crosses II, V and XXI (all grown under the same conditions in adjacent areas) show marked differences. In Composite Crosses V and XXI the frequency of resistant individuals remained below 30% over all generations examined including those [F_{32} in CC V (1973) and F_{16} in CC XXI (1975)] which overlapped in real time with the forty-fifth (1973) and forty-seventh (1975) generations of Composite Cross II (Figure 7.7; Jackson *et al.*, 1978). At the same time, studies of the pathogen in California have shown large changes in racial composition over time (Jackson, 1979). This further reduces the possibility that resistance to any particular race(s) would accumulate to the extent found in Composite Cross II solely in response to pathogen-induced selective pressures. In fact, considerable evidence has been obtained from a series of electrophoretic studies of Composite Crosses II and V (Clegg *et al.*, 1972; Weir, Allard & Kahler, 1972, 1974) to show that striking gametic disequilibria rapidly develop between loci within these populations. This non-random association of genes results in marked changes in the frequency of a whole range of alleles as blocks of genes respond to selection.

Overall, it seems that rather than representing an example of long-term, directional selection for resistance, the changes in frequency of resistance alleles detected in Composite Cross II are a reflection of the association of these alleles with linkage blocks which together respond to selection (Jackson *et al.*, 1978). Paradoxically, however, these data do indicate that when selection for disease resistance is sufficient to induce genotypic changes in populations of predominantly selfing plants, the effects of this selection are likely to ramify throughout the genome. Obviously, in agamospermous species widespread changes in seemingly unrelated characters are the inevitable consequence of effective selection for disease resistance. In Australian populations of *Chondrilla juncea*, selection against the susceptible narrow-leaved form has led to an increase not only in the frequency of resistance in mixed populations but also in the frequency of a wide range of electrophoretic and morphological markers (Burdon, Marshall & Groves, 1980; Burdon *et al.* 1981).

Pathogens and the evolution of sexual reproduction

A central concern in any consideration of the evolution of sex must be just what are the selective forces responsible for the maintenance of sexual reproduction in the face of the known costs involved in the production of males (Maynard Smith, 1978). In fact, the short-term advantages of parthenogenesis appear so great that a real dilemma is posed by the failure of parthenogenetic varieties to replace sexual ones. Solutions to this problem have been sought in arguments concerning the importance of sexual recombination in breaking down linkage disequilibria developed through random drift (Williams, 1975; Felsenstein & Yokoyama, 1976) or in the flexibility that recombination confers on populations in environments in which selection varies dramatically over time (Maynard Smith, 1971, 1978; Charlesworth, 1976). While these features are theoretically capable of explaining the continued occurrence of sexual recombination, biologically realistic scenarios under which the latter explanation may be sufficiently important have been difficult to provide (Maynard Smith, 1978). It is in this context that a number of workers have proposed recently that pests, parasites and pathogens, through the selective pressure they can exert on host populations, may in fact be the main driving force responsible for the evolution and maintenance of sexual reproduction in plant and animal species (Levin, 1975; Jaenike, 1978; Hamilton, 1980, 1982; Rice, 1983).

In particular, Hamilton (1980, 1982) has developed a series of gene-for-gene host–pathogen models involving one or two loci to study this possibility. He showed that, in addition to cyclic or chaotic fluctuations in host numbers and genotypes being a possible outcome of frequency-dependent selection, hosts which reproduce sexually may have a mean fitness which more than offsets the 'two-fold' cost of meiosis. These models have been critically assessed by May & Anderson (1983) who used differently derived fitness functions to show that both these models, and one incorporating density-dependent selection, produced predictions similar to Hamilton's only if pathogen infection was almost invariably fatal (< 0.001 survival rate). However, May and Anderson did point out that yet other fitness functions or complications involving more complex multilocus interactions and linkage disequilibrium might result in greater support for the proposed role for pathogens in the evolution of sexual reproduction.

Indeed, extensive reviews by Levin (1975), Clarke (1976) and Rice (1983) have covered a wide range of empirical examples and situations which provide some circumstantial support for this idea. Levin persuasively argued that an open recombination system would be of great significance to plant species suffering intense pathogen pressure. In such circumstances, populations of species which rely on agamospermy or other forms of asexual reproduction would, by virtue of their greater uniformity, be more vulnerable to disease and pest outbreaks than would those relying on sexual reproduction. Empirical support for this suggestion has come from a comprehensive data set developed from a review of literature concerning the biological control of weedy plant species (Burdon & Marshall, 1981*b*). Even after various potential biases involving the control of many closely related species and of the same species at many locations were eliminated, asexually reproducing species were controlled in 75% of cases in comparison with a 33% success rate for sexually reproducing hosts. In a further novel twist, Clarke (1976) has suggested that the frequency-dependent selection pressure mounted by pests and pathogens is largely responsible for the diversity of protein polymorphisms found in natural populations.

It has not been my intention to review here the substantial body of circumstantial evidence available which supports the hypothesis that parasites have played and continue to play an important role in the evolution and maintenance of sexual recombination in plant and animal species. However, the basic plausibility of this line of argument does emphasize the central role that pathogenic organisms potentially have in influencing the ecology and evolution of individual taxa and the composition of the plant communities in which they occur.

8

The effect of host plants on the population genetic structure of pathogens

The reciprocal interactions that are part of co-evolutionary models of resistance in the host and virulence in the pathogen are virtually indivisible. Increases or decreases in the frequency of resistance genes in the host population only make sense when considered in light of changes in the pathogen population. Similarly, fluctuations in the genetic structure of pathogen populations are best interpreted in terms of changes in that of the host.

Because this book is primarily about plant populations and the ways in which they may be affected by their pathogens, it was logical to consider the effects that pathogens have on the genetic structure of their hosts before the reciprocal interaction. For these reasons, theoretical models of host–pathogen interactions at a population level were considered in detail in Chapter 7. In the current chapter relevant aspects of the expectations and predictions of these models are reviewed briefly together with a number of models of pathogen development in agricultural mixtures. This is complemented by a consideration of some practical examples of the effects of host plants on the genetic structure of pathogen populations.

Theoretical models

Pathogen evolution in general host–pathogen models

The host–pathogen models developed by Jayakar (1970) and Leonard (1969*a*, 1977) (reviewed in Chapter 7) make similar predictions concerning the genetic structure of pathogen populations. Both models consider the effects of selection at a single locus for virulence in the pathogen and resistance in the host. This gives four possible combinations of host and pathogen, *viz*. susceptible host–avirulent pathogen, susceptible host–virulent pathogen, resistant host–avirulent pathogen, and resistant host–virulent pathogen. In both models different fitness values are assigned to host and pathogen in each of these combinations. Those for Leonard's model are detailed here.

The fitness values of pathogen genotypes in particular host–pathogen combinations are designed relative to 1.0 for the avirulent pathogen attacking a susceptible host (Table 7.2). On the resistant host, the fitness of the avirulent pathogen genotypes is $1\text{-}t$, where t represents the efficiency of that resistance gene in reducing reproduction of the pathogen. The fitness of the virulent pathogen genotype on the susceptible host is also assumed to be less than 1.0. In this case, fitness is reduced by a value k which represents the cost of virulence. In the fourth host–pathogen combination, that of a virulent pathogen and a resistant host, allowance (a) is made for the possibility that the pathogen may be more fecund ($a > 0$) when its virulence gene encounters the complementary resistance gene in the host (Leonard, 1977). The fitness of the virulent pathogen genotype in this combination is then $1-k+a$. Finally, when the two pathogen genotypes occur together in a mixed host population, the fitness of each is a product of the frequency of each host genotype and the fitness of the pathogen on that host. As Leonard (in press) has pointed out, the frequency of resistant plants that would provide for an equilibrium between the two pathogen genotypes varies according to the values assigned to the parameters a, k and t. Of these, k, the cost of virulence, has the most significant effect. If avirulent and virulent pathogen genotypes are equally fit on susceptible hosts (that is $k = 0$) then the virulent genotype eventually will totally exclude the avirulent one from the population. Both Leonard (1977) and Jayakar (1970) found that a polymorphic pathogen population could not be maintained without a fitness cost associated with virulence. Other models that deal with the dynamics of the genetic structure of pathogen populations make similar predictions (Person *et al.*, 1976; Groth & Person, 1977).

Pathogen evolution in multiline models

The basic assumptions underlying the models of pathogen evolution in mixtures and multilines have been detailed in Chapter 7. Despite these simplifications the predictions of the models are in good general agreement with those of more generalized host–pathogen interactions. Like the latter interactions, features that have been shown to be critical in determining the final outcome in any particular crop–pathogen situation are the level of selection against unnecessary genes for virulence (equivalent to Leonard's k) and the number of host genotypes in the mixture (Groth, 1976; Barrett & Wolfe, 1978; Marshall & Pryor, 1978). Moreover, high selection coefficients against unnecessary virulence usually have to be invoked ($k > 0.4$) to prevent the dominance of 'super-races' – pathogen genotypes that are virulent on all components of the mixture. Further modifications of these models have allowed for the effects of incomplete suppression of sporulation on resistant host lines (Marshall & Burdon, 1981) and of selection at different stages of the pathogen's life cycle (Marshall, Burdon & Muller, 1986). These features may significantly reduce the likelihood of the evolution of pathogen races with complex combinations of virulence.

One common feature of all models is that they deal exclusively with the genetic consequences, for the pathogen, of attacking a mixed host population. They make no attempt to integrate changes in the genetic structure of the pathogen population with changes in its size. An alternative approach, which potentially should provide greater realism, is to examine the relationship between evolution in the pathogen population and the epidemiology of the disease itself. This has been done by Barrett (1980).

The conclusions that Barrett draws from his model (Barrett, 1980, Barrett & Wolfe, 1980) are not inconsistent with those of the purely genetic models. However, the overall behaviour of the pathogen population is shown to be influenced markedly by features such as the pattern of spore dispersal within host stands. In the purely genetic models, pathogen spores are distributed randomly over the entire host population to initiate each new generation. This is, of course, a gross simplification. Spore dispersal patterns are typically characterized by inverse power relations between spore numbers and distance from the inoculum source (see Chapter 3). Barrett has accommodated this by partitioning the spores produced between those that enter an aerial pool from which they are distributed at random (ϕ) and those that reinfect the plant on which they were produced ($1-\phi$). In an equi-proportioned three-component mixture, when the value of ϕ is low ($\phi = 0.1$) most of the spores produced on a given host stay on that individual. As a result, simple pathogen genotypes carrying only one virulence gene dominate the population (Figure 8.1). However, as ϕ rises (for example, $\phi = 0.5$) and hence the proportion of the total spore production that is distributed at random also rises, complex pathogen genotypes with three genes for virulence rapidly become the most dominant class of individuals.

In any of these situations, the relative frequencies of the four possible classes of pathogen races (those with zero, one, two or three virulence genes) is also dependent on the number of generations that have elapsed since the establishment of the population. Even where simple races finally predominate (Figure 8.1a), pathogen genotypes carrying two or three

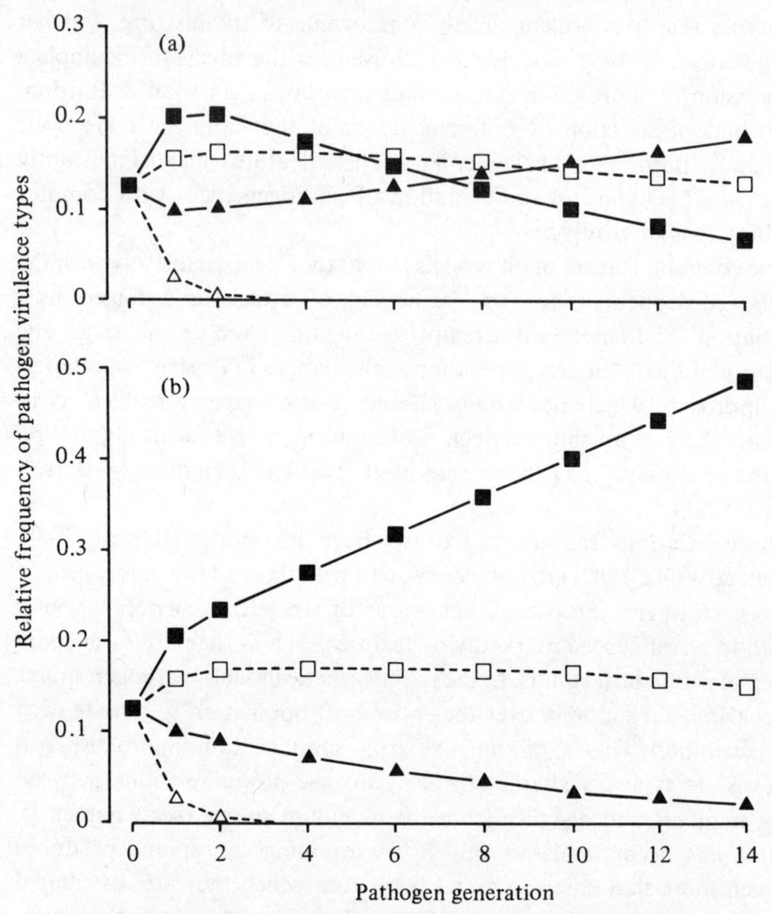

Figure 8.1 Relative frequencies of pathogen races carrying zero (△), one (▲), two (□) and three (■) virulence genes in a three-component host mixture simulated by the Barrett model. (a) and (b) differ in the values given to ϕ, the proportion of spores leaving the parent plant: (a) $\phi = 0.1$; (b) $\phi = 0.5$. (Redrawn from Barrett, 1980.)

virulence genes may be more common during the first few generations after establishment. In fact, a constant feature of this model is the initial rise in frequency that occurs for all pathogen genotypes capable of attacking more than one component of the mixture. This is perhaps not surprising as such genotypes are inevitably at an overall advantage during the initial infection of new host stands. At this time, spores are entering the stand from some distance away and will be distributed at random. Those capable of attacking only one of the three components will clearly be worse off than those able to attack two or three.

A common feature of all the theoretical models considered above is the assumption of a fitness cost (k) associated with virulence. Without such costs, a balance is not achieved between the different genotypes in the pathogen population. The race with the most complex virulence becomes fixed. Depending upon the particular argument being developed, the value of k, needed to achieve a polymorphic pathogen population, varies widely both within and between the different models. However, as the models incorporate more refinements that might be expected to affect real world interactions between hosts and pathogens, the magnitude of the values of k needed to prevent the dominance of a super-race fall. Indeed, in the simulations shown in Figure 8.1, $k = 0.1$.

Hopefully, the next generation of theoretical models will continue this trend towards increasing realism and, from the point of view of host–pathogen interactions in non-agricultural communities, will address the consequences of the patchy distribution of host plants. In the meantime, we turn here to a consideration of the performance of individual pathogen genotypes.

Empirical evidence

The performance of individual pathogen genotypes

The cost of virulence

The theoretical models considered here all rely on a fitness cost to be associated with virulence in the pathogen. What evidence supports this idea?

The frequencies of particular genes for virulence in pathogen populations may decline following reductions in the frequency of the corresponding resistance genes and this has often been advanced as an example of selection against unnecessary virulence* (van der Plank, 1968). Such field data must be interpreted with great care (see later). Race surveys often combine results from different sub-populations subject to different selection pressures. Furthermore, if the pathogens concerned rarely, or never, pass through a sexual cycle, the observed frequency changes may reflect no more than the inevitable decline in one clonal line as its food resource diminishes. For similar reasons the continued presence of unnecessary genes for virulence at high frequencies is not conclusive proof of the lack of a fitness cost. These may be retained: (1) as a result of active selection on other wild hosts in the same region; (2) through tight linkage to other

* van der Plank uses the term 'stabilizing selection' to describe selection against unnecessary virulence. As this term has a different, long-standing and widely accepted use in genetics it will not be used here.

genes vital for survival; or (3) in asexually reproducing populations, by a lack of sexual reassortment. However, if these potential problems are ignored or are not relevant to the particular pathogen in question, actual measures of the fitness costs can be obtained. This has been done for two host–pathogen combinations by Grant & Archer (1983). The cost of unnecessary virulence at the *Sr6* locus in the Australian *Puccinia graminis tritici* (wheat stem rust) population was estimated at between 4.0% and 5.1%. A similar estimate for the *Mla6* locus in the British *Erysiphe graminis hordei* (powdery mildew) population was between 5.4% and 6.1%.

Laboratory and glasshouse experiments in which different isolates of a pathogen are grown together provide a second source of data concerning the effects of unnecessary genes for virulence on pathogen fitness. Many such experiments have been carried out and these have been reviewed on several occasions (van der Plank, 1968; Nelson, 1973; Crill, 1977; Parlevliet, 1981). The available data are of variable quality, confusing and generally leave many questions unanswered. A comprehensive review of this literature is not appropriate here. Instead, some of the better examples of experimental approaches to this question are considered.

Many experiments investigating the potential cost of unnecessary virulence fail to take even the most elementary precautions to ensure that observed differences are due to variations in pathogenicity rather than to epistatic interactions within the genome of individual isolates. This may be attempted by restricting comparisons to isogenic lines of the pathogen that differ only at specific virulence loci. Alternatively, it may be achieved by monitoring changes in gene frequencies in large, heterogeneous pathogen populations. This should ensure that the genetic background of the virulence loci under study is randomized in the initial population.

Experiments involving competition between individual isolates of different pathogen races have frequently been reported. One experiment that largely avoided problems associated with differences in the background genome is that of Watson & Singh (1952). These workers compared, on a susceptible wheat variety, the performance of pairs of races of *P. graminis tritici* that differed only in pathogenicity at one or two virulence loci. In each combination, the more virulent race was believed to have arisen from the less virulent one by spontaneous mutation at the appropriate virulence locus. In all comparisons the isolate with the wider virulence was found to be less competitive than the one with a narrower host range. Leonard (1977) reanalysed these data and determined fitness costs associated with the possession of unnecessary virulence. These values were 0.42 for the gene for virulence on *Sr6* and 0.22 for that for virulence on *Sr11*.

However, even comparisons between apparently near-isogenic lines of a pathogen may be confounded by other effects. Ogle & Brown (1970) performed a similar competition experiment involving two Australian races of *P. graminis tritici* that differed in virulence at the *Sr9b* locus. Again the virulent race was believed to have arisen from the avirulent one by spontaneous mutation in the field. In a series of trials the more virulent one always performed better than the less virulent one. However, it was later shown that these races differed not only in virulence but also in aggressiveness (Ogle & Brown, 1971). The rate of pustule development, the ultimate size of pustules and the number of spores produced per pustule were all greater in infections involving the more virulent race. The genetic basis of these differences in aggressiveness were not determined.

Leonard (1969*b*) used an alternative approach to measuring the potential cost of virulence by studying selection in a large heterogeneous pathogen population. This population of *Puccinia graminis avenae* (oat stem rust) was established by inoculating plants of the susceptible oat variety 'Craig' with aeciospores collected directly from the pathogen's sexual host, *Berberis vulgaris*. Eight successive uredial (asexual) generations were then maintained on both 'Craig' and an additional variety. During this time, changes in pathogenicity in the population were monitored on oat plants with known genes for resistance. For every differential variety used, the frequency of components that were avirulent increased with each additional generation on the susceptible varieties. Overall, virulent races were 14–46% less fit on 'Craig' than were their avirulent counterparts (Leonard, 1969*b*). There was marked selection against races carrying unnecessary genes for virulence.

Using a similar type of approach, Alexander, Groth & Roelfs (1985) obtained results that could be interpreted as selection both for and against unnecessary virulence genes. These workers studied virulence changes in a heterogeneous population of *Uromyces appendiculatus* (bean rust) that was collected in the field and then maintained for five consecutive generations on the snap-bean variety 'Slimgreen'. At the end of the second and fifth generations, the population was assessed for the frequency of genes for virulence on the bean varieties 'US No. 3', 'B 1349', 'Early Gallatin' and 'Roma'. These values were compared with the virulence frequencies of the original field population (Figure 8.2). For the three varieties 'B 1349', 'Early Gallatin' and 'Roma' there was a steady decline in the frequency of virulence in the rust population. On 'US No. 3', however, there was a steady increase in virulence.

The specific causes of these changes in virulence were not determined,

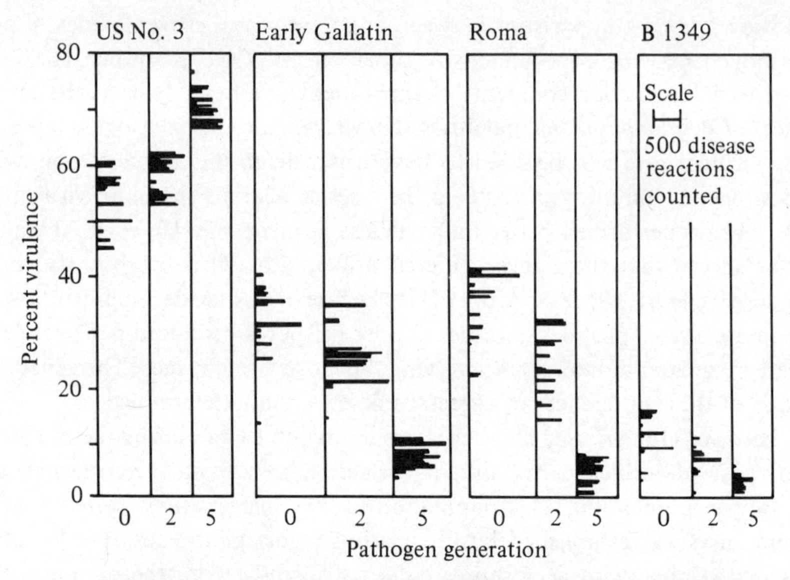

Figure 8.2 Frequency of pathogen phenotypes virulent on four differential bean lines for the original urediospore population of *Uromyces appendiculatus* and for populations maintained on the cultivar 'Slimgreen' for two and five asexual generations. (Redrawn from Alexander *et al.*, 1985.)

being due either to correlated responses to selection elsewhere in the genome or to the superior fitness of particular combinations of virulence genes (Alexander *et al.*, 1985). Certainly though, this study does underline the fact that differences in pathogenicity are not the sole criteria that determine the success or failure of individual pathogen genotypes.

These examples provide a vignette of a large number of studies that have attempted to measure the cost of unnecessary genes for virulence. It is unreasonable to expect all of these to be in total agreement. Fitness costs are unlikely to be the same for all virulence genes and may be affected by other factors like prevailing environmental conditions and the background genetic constitution of the individual pathogen isolate. However, the variability in results that sees genotypes advantaged, disadvantaged or unaffected by the possession of unnecessary virulence ensures that the controversy concerning the cost of virulence will continue!

Variations in the aggressiveness of isolates

While virulence for particular resistance genes is obviously important in determining the overall ability of individual pathogen isolates to attack particular host varieties, it has often been emphasized to the exclusion of all other variability. However, overall fitness of individual pathogen genotypes is determined by two distinct features of their performance – virulence and aggressiveness. The latter feature is a comparative measure of the degree of reproductive success that a virulent isolate has on a particular host genotype. Variations in this aspect of the fitness of individual isolates may have significant effects on the structure of pathogen populations. Furthermore, because aggressiveness is apparently inherited independently of virulence it is likely to have a confounding effect on assessments of the cost of virulence.

In many host–pathogen combinations there appears to be considerable intra-racial variation in features associated with aggressiveness. This is illustrated by differences that have been observed on a number of occasions in the growth of isolates of *Phytophthora infestans* (late blight) on potato. In one study, Caten (1974) collected two independent isolates of the same physiological race of *P. infestans* from each of three different potato varieties growing in the field. When these isolates were grown on tubers of all three varieties, four of the six pathogen isolates grew significantly faster on the host variety from which they were originally collected than on the other hosts. These differences in growth rate were as great as 20%. Results presented by Leonard (1969*b*) suggest that strains of *Puccinia graminis avenae* may also show similar levels of adaptation to specific host genotypes.

There is some suggestion that selection for increased aggressiveness as a result of repeated generations on the same host genotype may help to counter evolution towards greater virulence in pathogen populations attacking mixtures of different host genotypes. In such mixtures, the flexibility that some races achieve through the possession of multiple virulence may be at least partially offset by the greater fecundity of pathogen isolates that are more specialized and have become very closely matched to one particular host (Chin & Wolfe, 1984*b*) (see later).

The potential importance of selection for increased aggressiveness has largely been overlooked in theoretical models of the dynamics of pathogen populations. However, as a growing body of examples shows, it is a real phenomenon. In field situations where selective pressures operating against genes for virulence are likely to be much lower than those occurring in glasshouse experiments, selection for increased aggressiveness on particular host genotypes may contribute significantly to the final structure of pathogen populations. Chin & Wolfe's results concerning the behaviour of *Erysiphe graminis hordei* (powdery mildew) populations in mixtures of barley varieties suggest that similar processes may also occur in more natural mixed plant communities.

The virulence structure of pathogen populations

What is the virulence structure of typical pathogen populations? How do these populations respond to changes in their hosts? Ultimately, answers to these sorts of question will provide a clear indication of the degree of interaction that occurs between host and pathogen populations. Nearly all available data concern the structure of pathogen populations that attack agricultural crops. Unfortunately, unlike the majority of wild plant populations, these are usually genetically uniform and as such are far from even the simple mixed populations of many theoretical models. However, even data from these interactions are useful. They provide insight into both the effects that hosts may have on their pathogens and into the care that is needed in interpreting survey results.

Patterns of racial variation in agricultural plant pathogens

Virulence or race surveys are carried out routinely for a range of pathogens of agricultural crops. Such surveys are especially extensive for rust and mildew diseases of cereals (Wolfe & Minchin, 1976; Young & Prescott, 1977; Green, 1979; Luig, 1979) but have also included assessments of other pathogens like *Bremia lactucae* (downy mildew of lettuce; Gustafsson & Larsson, 1984), *Rhynchosporium secalis* (barley scald; Hansen & Magnus, 1973) and *Phytophthora infestans* (potato blight; Shattock, 1974).

Typically, such surveys find that pathogen populations are not monomorphic. Rather they are composed of a range of different races, although a small number of these may dominate the population. Race surveys of *Puccinia graminis tritici* (wheat stem rust) populations occurring in Australia provide representative examples of this pattern. During a survey of this pathogen in the 1973–74 growing season 36 different races were detected on wheat (Luig & Watson, 1977). The commonest of these races occurred with a frequency of 0.17 while the combined frequency of the three commonest was 0.42 (Figure 8.3*a*). By contrast, the 24 least common races made up less than one tenth of the population. Seven of these were represented by a single isolate only. Eight years later, the frequency distribution of different races in the same population was even more unbalanced (Burdon, Luig & Marshall, 1983). By that time, the pathogen population was totally dominated by three other races, the combined frequency of which exceeded 0.8 (Figure 8.3*b*). One of these races alone

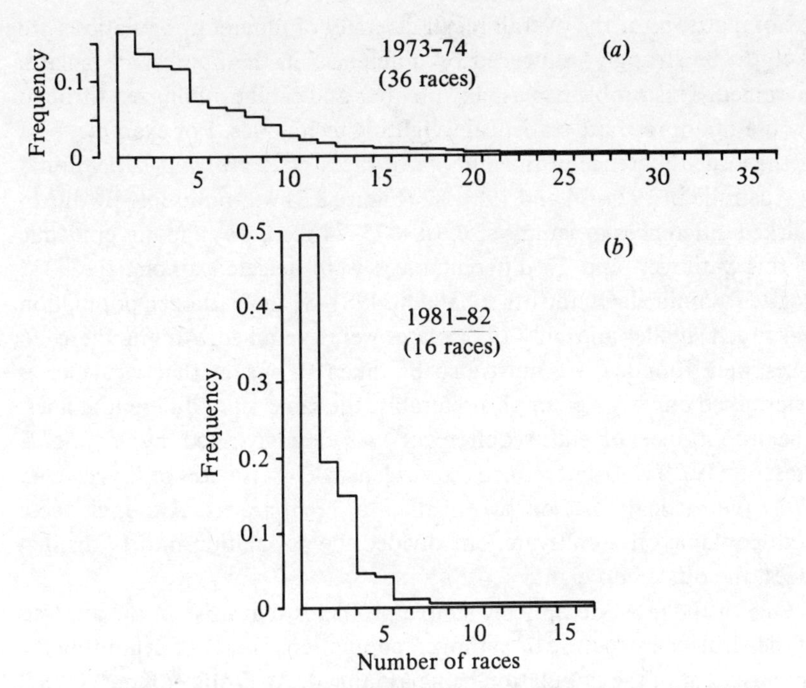

Figure 8.3 Frequency distributions of races of *Puccinia graminis tritici* detected in Australia in the 1973–74 and 1981–82 wheat growing seasons. (Derived from data of Luig & Watson, 1977; Burdon *et al.*, 1983.)

accounted for half of all isolates examined! In addition, none of the dominant races was the most virulent. Similar frequency distribution patterns have been detected in a wide range of other pathogens (for example, Groth & Roelfs, 1982).

Pitfalls in the interpretation of survey data

Pathogen race surveys like these provide a very generalized picture of the level of variation occurring in different populations, the presence of specific virulence phenotypes and the extent to which individual populations are dominated by a small fraction of the races that are present. However, more extensive analysis of such data is fraught with potential difficulties (Wolfe & Knott, 1982). These difficulties may be divided into two basic groups according to their ease of recognition. Widely recognized problems include the effects of variable sample sizes and differences in the differential sets used to determine virulence. Less obvious problems include features like the delimitation of the boundaries of a pathogen population and the mode of reproduction of the pathogen concerned.

Comparisons of the overall racial diversity of different populations are likely to be strongly influenced by differences in the number of isolates examined. This problem is usually obvious and can be minimized through the use of appropriate statistical weighting techniques. For example, part of the large difference in the number of races of *P. graminis tritici* found in Australia in 1973–74 and 1981–82 (Figure 8.3) was undoubtedly due to marked differences in sample size. In 1973–74 there was a major epidemic of this pathogen and 36 different races were detected among the 1317 isolates examined. On the other hand, in 1981–82 the pathogen population was much smaller and only 193 isolates were examined. Among these, 16 races were found. Care must also be taken to ensure that virulence is determined on very similar or, preferably, the same set of differential lines. The importance of this requirement was demonstrated by Young & Prescott (1977) who determined the virulence of 647 isolates of *P. recondita tritici* (wheat leaf rust) on two distinct differential sets. Although these both contained five cultivars, one divided the population into 17 distinct races, the other into eight.

One of the most deceptively simple problems that arise in the analysis of the virulence structure of pathogen populations is that of delimiting the actual extent of the population being examined. As Wolfe & Knott (1982) have pointed out, pathogen isolates for race surveys can be collected in a variety of ways. Each of these places restrictions on the extent of analysis that is subsequently possible. In fact, most published virulence survey data do not represent a collection of isolates gathered from a single population that has developed under a particular set of selective forces. Rather, they are the combined results of the uneven sampling of a number of sub-populations, occurring on a range of differential commercial cultivars, carrying a range of different resistance genes. Such sub-populations are likely to have been subject to quite different selective forces and may not have interacted with one another (Wolfe & Knott, 1982).

Analysis of the race survey data for *P. graminis tritici* in Australia in 1973–74 provides an example of the consequences of some of the problems raised by Wolfe & Knott. If data for the pathogen populations occurring in southern New South Wales and South Australia are extracted from that of the total Australian population, several differences appear (Figure 8.4). In both areas the combined frequency of the three commonest races exceeded 0.5 – a marked increase in dominance relative to the Australian population as a whole (Figure 8.3). Of greater importance, however, are the differences in the frequency of occurrence of specific pathogen races. These were most prominent for race 21-1,2 which was the commonest race

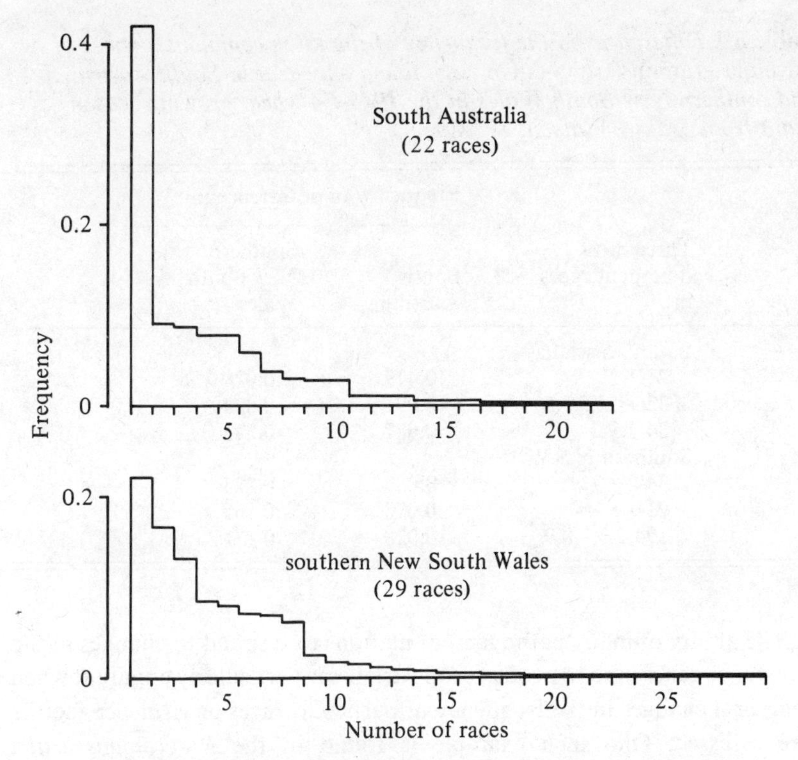

Figure 8.4 Frequency distributions of races of *Puccinia graminis tritici* detected in southern New South Wales and South Australia in 1973–74. (Derived from data of Luig & Watson, 1977.)

in South Australia where it occurred with a frequency of 0.42. In southern New South Wales the frequency of the same race was only 0.08 (Table 8.1). Similar, although less pronounced, differences were detected for the three races that dominated the *P. graminis tritici* population in southern New South Wales. The frequency of these races in South Australia was less than half that in southern New South Wales. It is most probable that these differences in the virulence of the predominant races reflect differences in the selective pressure exerted on the pathogen by the wheat cultivars grown in the two areas. In the early 1970s more than half the South Australian wheat acreage was sown to 'Halberd'. This variety carries the resistance genes *Sr6* and *Sr11* but is susceptible to race 21-1,2 which possesses the complementary virulence alleles. In southern New South Wales, on the other hand, the recommended wheat varieties 'Gamenya' (*Sr9b*), 'Robin' (*Sr9b*, *Sr11*) and 'Summit' (*Sr5*) have apparently conferred a selective advantage on race 34-2,3,7 which carries the appropriate virulence genes.

Table 8.1 *Differences in the frequency of the three commonest races of* Puccinia graminis tritici (*stem rust*) *found on wheat in South Australia and southern New South Wales in the 1973–74 wheat growing season (data from Luig & Watson, 1977)*

Three most dominant races in:	Frequency of occurrence in South Australia	Southern New South Wales
South Australia		
21-1,2	0.419	0.078
326-1,2,3,5,6	0.091	0.023
34-2,3,7	0.087	0.223
Southern N.S.W.		
34-2,3,7	0.087	0.223
21-2,5	0.078	0.165
194-2,3,7,8,9	0.028	0.131

The ability of many pathogen populations to respond to changes in the resistance genes used to protect crop varieties is particularly apparent when temporal changes in the frequency of particular races or virulence factors are assessed. One such example is found in the *P. graminis tritici* population occurring in northern New South Wales and Queensland (Luig & Watson, 1970). In the early 1950s wheat varieties grown in this area were protected only by single resistance genes (e.g., 'Gabo' (*Sr11*) and 'Gamenya' (*Sr9b*)). As these varieties were overcome by the pathogen, they were progressively replaced by others carrying simple (e.g., 'Tarsa' (*Sr6*, *Sr9b*)) and then, later, complex (e.g., 'Gamut' (*Sr6*, *Sr9b*, *Sr11*, *SrGt*) and 'Mendos' (*Sr7a*, *Sr11*, *Sr17*, *Sr36*)) combinations of resistance genes. In turn each of these varieties also became susceptible. This progressive step-wise evolution of the pathogen population resulted in the average number of virulence genes per pathogen race rising from 1.46 in the period 1954–58 to 3.18 in 1964–68 (Figure 8.5; Luig & Watson, 1970).

Fluctuations in the virulence structure of the *P. graminis avenae* (oat stem rust) population in western Canada provide a further example of the responsiveness of pathogen populations to changes in their hosts (Martens, McKenzie & Green, 1970). In the 1940s the frequency of virulence on the *Pg2* resistance gene rose rapidly as a result of the widespread cultivation of oat varieties that were protected by this gene alone. When the variety 'Rodney' carrying *Pg4* was released, it was resistant to the prevailing pathogen population and gained wide accept-

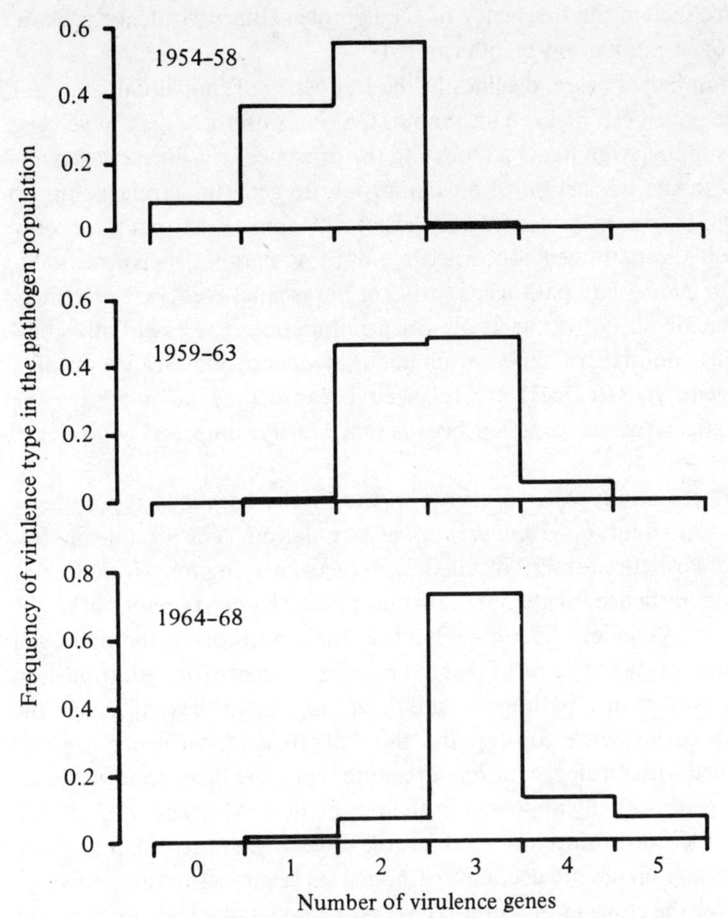

Figure 8.5 Frequency distributions of different virulence types in the *Puccinia graminis tritici* populations occurring in northern New South Wales over the period 1954–68. (Redrawn from Luig & Watson, 1970.)

ance with farmers. The established varieties carrying *Pg2* were rapidly replaced and a corresponding decline occurred in the frequency of the gene for virulence on *Pg2*. In the late 1950s the release of oat varieties with combined *Pg2* and *Pg4* resistance led to a resurgence in the frequency of the gene for virulence on *Pg2*.

In addition to illustrating the ability of pathogen populations to 'track' genetic changes in their host, the last example also highlights a second major problem that may arise in the interpretation of data from race surveys. This problem derives from the mode of reproduction of many plant pathogens and may be best summed up in a question. Following the replacement of a host variety protected by a specific resistance gene, how should declines in the frequency of the complementary virulence gene in the pathogen population be interpreted?

In a number of cases, declines in the frequency of individual genes for virulence have been marked and rapid. On occasion these may have been used to support arguments advocating the existence of a fitness cost associated with the possession of an inappropriate gene for virulence in the pathogen. However, as considered earlier, this line of reasoning can only be valid if the pathogen concerned regularly undergoes sexual recombination. In many plant pathogens sexual or parasexual events are extremely infrequent or do not occur at all. Such populations are essentially composed of a number of different clones that do not interact genetically. Whole genotypes (clones) are replaced because they do not have the appropriate virulence gene *not* because they carry unnecessary virulence genes.

This is essentially what occurred in the totally asexually reproducing western Canadian *P. graminis avenae* population. The decline in frequency of virulence for *Pg2* resulted from races carrying *Pg2* virulence not possessing virulence for the *Pg1* resistance gene that was present in newly released oat varieties. As a consequence, their position in the pathogen population was taken by races that did possess virulence for *Pg1*. In eastern Canada, where this pathogen undergoes an annual sexual cycle, the frequency of virulence for *Pg2* did not fall. Instead, virulence for *Pg2* recombined with virulence for *Pg1* to ensure that races possessing virulence for both genes rapidly appeared in the population (Martens *et al.*, 1970).

When the constraints imposed by the breeding system of individual pathogens are taken into account, pathogen survey data provide equivocal support for the concept of a fitness cost associated with virulence. Even in pathogen populations in which the sexual stage is common and no genes for race specific resistance have apparently been used, the frequency of genes for virulence may still be high. In a population of *Erysiphe graminis tritici* (wheat powdery mildew) occurring on wheat in southern Sweden, Leijerstam (1965, cited by Wolfe & Knott, 1982) found that the frequency of four unnecessary genes for virulence ranged from 0.11 to 0.65. This suggests that there was little selection acting against these genes. However, many agricultural plant pathogens also attack one or more wild hosts. It is always possible that the observed virulence frequencies reflect the effects of selection on these hosts.

The effect of agricultural mixtures on pathogen populations

For those interested in interactions between hosts and pathogens in non-agricultural systems, data collected from pathogen populations attack-

Table 8.2 *Percentages of races of* Erysiphe graminis hordei *(barley powdery mildew) with simple and complex virulences in pure and equi-proportioned three-component mixed stands of the barley cultivars 'Hassan', 'Midas' and 'Wing' (from Chin & Wolfe, 1984b)*

	Combined simple virulences	Complex virulences			
	H; M; W	H+W	H+M	M+W	H+M+W
Mean of pure stand	72.3	10.6	9.7	2.8	4.5
Mean of mixture (H:M:W)	61.6	12.8	5.2	8.1	12.2

ing varietal mixtures are of particular interest. Unfortunately, data of this nature are very limited. The most detailed study available is that of Chin & Wolfe (1984*b*) who monitored changes in the pathogenicity of populations of *Erysiphe graminis hordei* (powdery mildew) attacking mixtures of three barley varieties. These they compared with changes occurring in populations attacking pure stands of the component varieties. The response of the pathogen population to simultaneous selection by the different hosts was complex and difficult to predict. Although host mixtures often encouraged the gradual emergence of races with combined virulence, these did not necessarily replace simple races. The data obtained in one experiment provide an example of this variability in selection for races with complex combinations of virulence (Table 8.2). In the mixture, the relative frequency of races virulent on all three varieties ('Hassan', 'Midas' and 'Wing') and that of those virulent for the combination of 'Midas' and 'Wing' both increased. However, the relative frequency of the race virulent on 'Hassan' and 'Wing' showed little change while that of the third pair-wise combination ('Hassan' and 'Midas') actually fell. In both pure stands and the mixture, races with virulence for only one of the three varieties predominated. Other tests indicated that simple pathogen races possessing the appropriate virulence gene dominated pure stands by virtue of higher survival and reproductive rates. Presumably, however, in mixed host populations the observed lower fitness of complex races was compensated for by their increased flexibility. In some cases, the latter advantage apparently more than offset the disadvantage of lower fitness and resulted in selection for a complex race. In other situations, the advantage of flexibility was not as great and a simple race was favoured (Chin & Wolfe, 1984*b*).

The virulence structure of non-agricultural plant pathogens

Very little is known about the virulence structure of populations of non-agricultural plant pathogens. Certainly, as studies of *Colletotrichum gloeosporioides* (anthracnose) attacking *Stylosanthes guianensis* in Colombia (Miles & Lenné, 1984) and of *Erysiphe fischeri* (powdery mildew) attacking *Senecio vulgaris* in Britain (Clarke & Harry, 1979) have shown, a range of different virulence biotypes occur in wild pathogen populations. Unfortunately, neither of these studies examined a sufficient number of pathogen isolates to allow any consideration of the extent of diversity in the population or of its causes. Indeed, *such data currently do not exist for any wild host–pathogen system* in which the host range of the pathogen in question does not overlap onto agricultural crops.

Undoubtedly a part of the reason for this lack of data lies in problems associated with the determination of virulence responses. This can be a daunting task as the virulence characteristics of individual pathogen isolates cannot be determined until a suitable set of host lines, each differing in resistance, has been assembled (see Chapter 4). This alone is likely to take considerable time and effort. However, in the case of pathogens whose range includes both wild and agricultural plants, suitable sets of differential hosts may already be available.

In situations where resistant varieties of the crop have been grown in the field, studies of such host–pathogen interactions are of dubious value. It will not be possible to distinguish between the effects of the cultivated or the wild hosts on the pathogen population. On the other hand, if the agricultural host is uniformly susceptible, then study of the pathogen population on the wild host should provide useful information concerning the effects of wild plants on the population structure of their pathogens. The interaction occurring between the pathogen *Puccinia coronata* (crown rust) and the introduced wild oat species *Avena barbata*, *A. fatua* and *A. ludoviciana* in Australia, provides an example (Burdon *et al.*, 1983; Oates *et al.*, 1983).

The significant differences that occur between the resistance structure of wild oat populations growing in northern and southern New South Wales have been described in Chapter 7. The mean level of resistance of populations growing in the northern part of the state was greater than that of southern populations, as was the overall diversity within populations. When samples of *P. coronata* collected from wild oat plants growing throughout New South Wales were analysed on a similar geographical basis, some interesting parallels with the host resistance data emerged. Over the 5-year sampling period, an annual average of 9.0 races of the

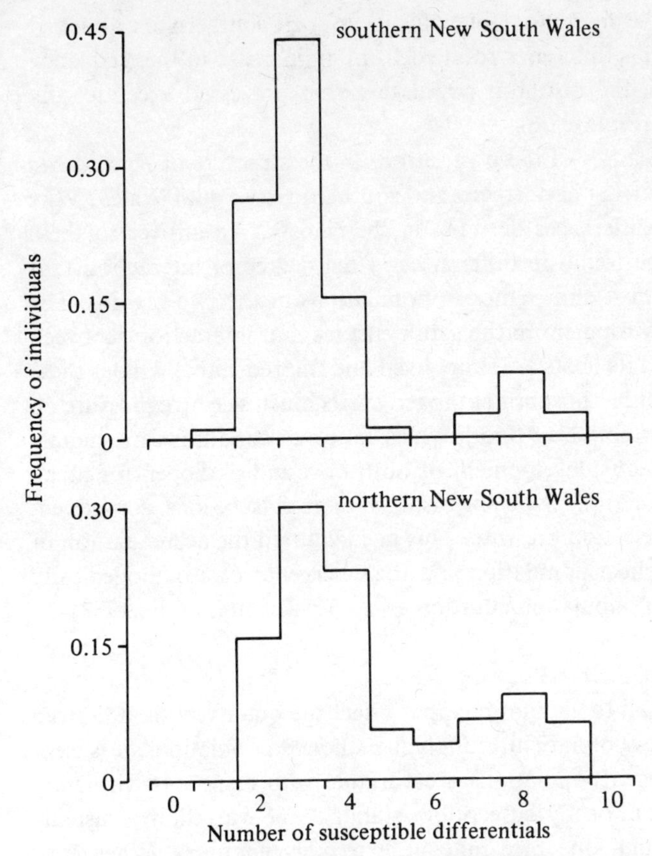

Figure 8.6 Differences in the frequency distributions of virulence types in the *Puccinia coronata* population occurring in southern and northern New South Wales, averaged over the period 1975–79. (From Oates *et al.*, 1983.)

pathogen was detected in the north as opposed to 5.8 in the south. Even when these values were corrected for variations in sample size, significant differences remained between the racial diversity of the northern and southern populations of *P. coronata* (Oates *et al.*, 1983).

Not all races of a pathogen are equal with respect to pathogenicity. Some races possess more virulence alleles than others and, as a consequence, are able to attack a wider range of host genotypes. To allow for this possibility, the pathogenicity of isolates of *P. coronata* was measured by the number of differential hosts they were able to attack. For individual races this ranged between zero and nine. Once again disparities were found between the populations of *P. coronata* occurring on wild oats in northern and southern New South Wales. The average northern isolate was virulent on

4.4 members of the differential set, while the average southern one attacked 3.5 host lines. This difference resulted from an increase in the frequency of individuals in the northern population that possessed exceptionally broad virulence (Figure 8.6).

What are the causes of these variations in the structure of *P. coronata* populations occurring in northern and southern New South Wales? Why do they mirror differences detected in their hosts? An answer to these questions may be found in differences in the degree of interaction that occurs between host and pathogen populations in the two regions. The nature of the environment in the south ensures that interactions between the pathogen and its hosts are short-lived and unpredictable. Under these circumstances neither host nor pathogen exerts much selective pressure on its partner. In northern New South Wales, however, conditions are suitable for the simultaneous development of both host and pathogen for much longer periods of time. Not surprisingly there has been a consequent greater mutual selection pressure. This has favoured the accumulation of resistance in the host population and the emergence of a pathogenically diverse pathogen population (Burdon *et al.*, 1983; Oates *et al.*, 1983).

Concluding remarks

Despite all the limitations that affect the quality of data derived from racial surveys of agricultural plant pathogen populations, it is clear that most pathogen populations are variable with respect to virulence. Even in large, uniformly susceptible stands some variation is usually detectable although one race may be extremely common. When host genotypes are mixed together, either as fields of different pure stands or as varietal mixtures, the number of different pathogen genotypes may be very much larger but a handful of these is still likely to dominate the population. Furthermore, data from racial surveys clearly indicate that host plant populations are able to exert strong selective pressure on those of their pathogens.

While these perceptions are derived from agricultural host–pathogen systems, the little evidence available from non-agricultural ones tends to suggest the same *general* features are applicable. That is, pathogen populations are likely to be variable with respect to virulence and are likely to respond to selective pressures exerted by their hosts. However, whether they will show the dramatic spatial or temporal phenotypic changes in virulence that have been observed in pathogens of agricultural crops is still a matter for speculation.

As has previously been remarked upon, wild plant populations are typically fragmented into many small patches. The effect this fragmentation has on the genetic structure of pathogen populations is difficult to predict. Although information concerning the resistance structure of localized patches of host is very limited, that which is available suggests that considerable variation often occurs (see Chapter 7). What effect is this likely to have on pathogen variability? Certainly, during the initial establishment phase, virulent pathogen isolates will be favoured over less virulent ones. However, the diverse host population is also more likely to favour selectively a range of virulent phenotypes than is a genetically uniform host population. Potentially this will lead to at least a limited range of pathogen variability within each sub-population associated with a host patch. The extent to which this will occur will be determined by a range of factors including: (1) the size and genetic structure of host patches; (2) the dispersal powers of the pathogen both within and between patches; and (3) the average number of pathogen generations that such individual host and pathogen sub-population associations last. These factors will also affect the extent of spatial differences between individual sub-populations of the pathogen.

Dramatic temporal fluctuations in the racial structure of agricultural plant pathogen populations are largely caused by rapid man-made changes in the particular crop varieties being grown in an area. Such rapid replacements do not occur in wild plant populations. However, in such systems, marked fluctuations in the genetic structure of pathogen populations are possible. These may occur particularly as a result of random fluctuations as population sizes are reduced to a minimum during unfavourable conditions.

As more is learnt about the genetic structure of populations of pathogens that attack wild plants it will be possible to gain a more precise picture of their temporal and spatial stability and of how they interact with populations of their hosts.

9

Further study

The main features apparent in reviewing the preceding chapters are the many yawning gaps in our knowledge of the interactions between plants and their pathogens in non-agricultural communities. Currently, our understanding of the ecological role of plant pathogens is rudimentary. There are virtually no areas for which there is an adequate range of experimental examples, while for many topics our 'knowledge' is little more than an extrapolation of observations made in highly specialized agricultural systems.

Unfortunately, wild host–pathogen systems have stimulated little interest among population biologists. The small and frequently insignificant visual appearance of pathogens has all too often been taken to represent a full measure of their ecological relevance. Indeed, as Price (1980) stated so eloquently when considering the neglect of parasites in general, 'Visually stimulating organisms, the large, the colourful, the active, the aggressive, command our attention while the secretive and insidious remain largely ignored *without regard for their relative importance*' (my italics). Nowhere is this statement more true than for pathogens of wild plant communities. Despite this neglect, sufficient evidence is already available to indicate at least some of the far-reaching effects that pathogens have on the dynamics and genetic structure of individual plant populations and the communities in which they occur. The challenge ahead is to unite these observations into a cohesive framework that explains the multitudinous effects of pathogens on plants and of plants on pathogens.

Ultimately, a full understanding of the role of pathogens in the ecology of plants will only come with detailed analyses of the dynamics of changes occurring in both host and pathogen populations as they respond to each other. At present our knowledge in this area is fragmentary. Of the role that individual pathogens play in the frequently documented seasonal and yearly fluctuations in the size, density and local patterns of distribution of individual plant populations we know little. Our knowledge of the genetic structure of host populations is similarly restricted to single snap-shots in time that are rarely, if ever, repeated in the same population. On the other side of the host–pathogen equation, our knowledge of the size and genetic structure of pathogen populations is almost exclusively derived from agriculturally based studies that provide a composite picture of entire populations summed over whole seasons.

Despite the magnitude of the task, studies that simultaneously follow changes in all these features will ultimately be needed if a complete understanding of the role of pathogens in the population biology of plants is to be achieved. By following seasonal and yearly fluxes in populations of both hosts and pathogens a true measure of the interactive effects of the two populations will become apparent. Certainly, though, detailed studies of limited parts of host–pathogen associations will continue to make significant contributions and should receive enthusiastic support. In particular, such studies may be able to provide more immediate answers to some of the pressing questions concerning topics such as: (1) the magnitude of the hypothesized costs associated with resistance in the host and virulence in the pathogen; (2) the relative importance of race specific and race non-specific resistance mechanisms in host populations; and (3) the effects that patchy host distributions have on the size and structure of pathogen populations. The last topic is also ripe for the attention of theoretical modellers – hopefully restrained to some semblance of biological reality by the active involvement of population biologists and pathologists! In providing answers to these sorts of questions, studies of host–pathogen interactions in natural plant communities will also make significant contributions to a wide range of topics that currently hold centre stage in their respective disciplines. Basic questions concerning the occurrence of co-evolutionary interactions in nature, the causes and controls of community diversity and stability, and the development and maintenance of complex polymorphisms can all be addressed in appropriate host–pathogen associations.

In many places the arguments developed in this book have relied upon examples derived from studies of agricultural systems. It is my belief that, in turn, the results of studies of wild plant–pathogen interactions may be exploited in the development of efficient weed and disease control strategies for use in agricultural ecosystems.

The biological control of weedy plant species through the use of natural enemies is a well established approach to weed control. To date most biological control programmes have used herbivorous insects as control agents and have generally ignored the possibility that the success of a programme may be affected by the genetic structure of the target weed (Burdon & Marshall, 1981*b*). However, as more interest is shown in the use of plant pathogens as control agents, a greater knowledge of the genetic consequences of the highly specific nature of interactions between pathogens and their hosts will become increasingly important. In particular, information concerning (1) the frequency and number of resistance genes encountered in marginal *versus* central plant populations, (2) the effect of the reproductive system of plants on the distribution of resistance genes within individual populations, and (3) the long-term fate of resistance genes in populations unchallenged by appropriate pathogens, will all materially assist in the development of control strategies that maximize the possibilities of success. These are all topics that are an integral part of the studies discussed earlier.

Studies of wild host–pathogen associations are also relevant to the development of strategies for the control of disease in agricultural crops. However, the recent controversy concerning the relevance of the gene-for-gene theory to non-agricultural plant–pathogen associations highlights the paucity of our knowledge concerning the relative importance of disease resistance mechanisms in such interactions. Certainly, in wild host–pathogen associations disease levels are, at least partially, held in check by a range of ecological and phenological phenomena that have little or no importance in agricultural systems. However, race specific and race non-specific resistance mechanisms are also present, although the relative importance of these is unclear. Traditionally, breeding for disease resistance in agriculture has relied heavily on race specific resistance to protect crops from pathogen attack. Such resistance, which is typically associated with gene-for-gene systems, has been assumed by some workers to be of little or no importance in wild systems. This assumed difference has been used to explain the apparent instability of disease control strategies in agriculture. In fact, however, where sufficiently detailed genetic analyses have been carried out in wild plant populations, race specific resistance mechanisms *do* seem to play a significant, although not exclusive role, in protecting individual plants against disease. Further detailed studies of the ways in which plants in natural populations are protected against disease may well stimulate further developments in strategies for the protection of agricultural crops.

This is particularly well illustrated by the relative recent trend towards the use of varietal mixtures and multilines in agriculture. Mixtures have often been justified by the claim that they mimic the diversity found in natural plant communities and hence will suffer less disease. However, these claims in themselves are poorly documented and a more profound knowledge of (1) the genetic structure of both wild host and pathogen populations, (2) the relative importance of different resistance mechanisms in the host population, and (3) the epidemiological behaviour of the pathogen population, will all provide significant insight into the long-term viability of such mixture-based disease control strategies for agriculture.

In the current era of limited research funding and greater demands for public accountability such applied aspects of wild host–pathogen studies are likely to become increasingly important.

Glossary of plant pathological terms

This glossary provides definitions for most of the pathological terms used in the text which may be unfamiliar to population biologists. These have been formulated with considerable assistance from the invaluable lists of definitions published by the Commonwealth Mycological Institute (1968) and the Federation of British Plant Pathologists (1973).

Aggressiveness This is a comparative measure of the degree of success that a virulent pathogen isolate has on a particular host genotype. It is assessed in terms of features such as lesion or pustule size, the number of spores produced and the rate of growth of the pathogen.

Allo-infection The process by which a host plant is infected by inoculum produced by lesions occurring on other host individuals (cf. auto-infection).

Autoecious Used of a pathogen to indicate the ability to complete the entire life cycle on a single host species. For example, *Melampsora lini* (the causal agent of flax rust) on *Linum usitatissmum* (cf. heteroecious).

Auto-infection The process by which a host plant is infected by inoculum produced elsewhere on the same individual (cf. allo-infection).

Biotrophic pathogen An obligate pathogen that is entirely dependent upon another living organism as a source of nutrients.

Differential hosts A set of host lines used in the identification of specific physiological races of a pathogen. Ideally, each host line should carry only one resistance gene.

Disease avoidance A relative lack of disease on a potentially susceptible host that results from the occurrence of one or more genetically controlled mechanisms in the host. These mechanisms prevent physical contact between host and pathogen. An example of this phenomenon is the closed flower habit of many cereals that prevents infection by *Claviceps purpurea* (the causal agent of ergot) (cf. disease escape).

Disease escape A relative lack of disease resulting from the failure of a susceptible host and pathogen to come into physical contact although both are present in the same environment at the same time. Disease escape is a reflection of a simple spatial phenomenon (cf. disease avoidance).

Focus A site of local concentration of diseased plants or disease lesions usually around a primary source of infection. The size and shape of foci tends to influence the pattern of further spread of the disease.

Gene-for-gene hypothesis The concept that for each gene that determines resistance in the host population there is a corresponding gene for virulence in the pathogen population with which it specifically interacts.

Heteroecious Used of a pathogen to indicate a requirement for two unrelated host species to complete the entire life cycle. For example, *Puccinia graminis* f.sp. *tritici* (the causal agent of wheat stem rust), whose sexual stages occur on *Berberis* spp. while the asexual ones occur on *Triticum* spp. (cf. autoecious).

Hypersensitive response An active resistance mechanism in which pathogen colonization of the host is prevented by the rapid death of cells around the point of infection. Typically this response is characterized by macroscopically visible, dark flecks.

Incubation period The time elapsed between infection and the first macroscopic signs of disease (cf. latent period).

Infection, primary The initial infection(s) that result from the successful transmission of inoculum from some external source to the host population in question. These primary infections are the future centre of disease foci from which inoculum is dispersed to other plants in the stand (cf. secondary infection).

Infection, secondary Infections which result from the reproduction and subsequent dispersal of inoculum from primary infections within host stands. Secondary infections will lead to a flattening of primary disease dispersal gradients.

Infection type The macroscopic appearance of disease lesions at the conclusion of the latent period. Typically these symptoms are assessed on a semi-continuous scale that ranges from immunity to full susceptibility. The major divisions of such a scale for *Puccinia graminis tritici* on wheat are shown in Figure 4.1.

Infectious period The period of sporulation or production of propagules.

Latent period The time elapsed between infection and the production of new pathogen propagules. (This definition follows that of van der Plank (1963) that has been generally adopted in epidemiology. It is in conflict with the proposals of the Federation of British Plant Pathologists (1973).)

Multiline Different lines of the one cultivar mixed together to form a composite variety. Multilines are developed by recurrent backcrossing of suitable lines to a common parent. As a result, all lines are agronomically very similar, differing in little more than the resistance genes they carry. Their genetic diversity per component line is less than that of varietal mixtures (cf. varietal mixture).

Pathogenicity The characteristic of being able to cause disease.

Physiological race *See* race.

Race A taxon of pathogens characterized by specialization to different cultivars of one host species. As used in this book, isolates of the one race would be characterized by identical patterns of virulence and avirulence.

Resistance The ability of a host plant to retard the activity of a pathogenic organism. Resistance in plants has been described by a range of terms which, while covering the same set of phenomena, do not correspond precisely. Terms commonly used in this book are described below.

Resistance, active Resistance resulting from host reactions occurring in response to the presence of a pathogen or its metabolites (cf. passive resistance.)

Resistance, adult plant Resistance detected at the post-seedling stage of development. This may be assessed at a single time or, as is frequently the case, over the course of a disease epidemic.

Resistance, horizontal Resistance which is effective against all races of a pathogen (van der Plank, 1963). This is equivalent to race non-specific resistance. The latter term is used in this book as van der Plank's definition of horizontal and vertical resistance is confounded by an inaccurate assessment of their epidemiological consequences (cf. vertical resistance).

Resistance, passive Resistance which is due to innate qualities possessed by the host prior to being attacked by a pathogen (for example, thick cuticles). It is probable that such resistance will be expressed against all races of a pathogen.

Resistance, race non-specific Resistance that is uniformly expressed against all races of a pathogen. It is commonly, but not necessarily under the control of many genes.

Resistance, race-specific Resistance that is recognized by a differential interaction between host and pathogen genotypes. It is effective against some races of a pathogen but not against others. Race-specific resistance usually relates to cases of resistance conditioned by one or a few genes.

Resistance, seedling Resistance detected during the early stages of plant development. In studies involving pathogens responsible for rust and mildew diseases of cereals, this is assessed when the first leaf is fully expanded. Resistance expressed at this stage of development may or may not be expressed in post-seedling stages.

Resistance, vertical Resistance which is effective against some, but not other, races of a pathogen. It is equivalent to race-specific resistance (cf. horizontal resistance).

Tolerance A measure of the relative ranking of the performance of two or more lines of a host species in the presence of disease, expressed as a fraction of their respective performance in its absence.

Varietal mixture A mixture of two or more different crop cultivars. Individual cultivars are generally chosen for inclusion in such mixtures on the basis of differences in their resistance to specific races of particular pathogens. However, because the cultivars used in such mixtures frequently have quite different pedigrees, they may also show variability in their response to other environmental factors including other pathogens. Varietal mixtures have a greater overall level of genetic diversity than do multilines with the same number of component lines (cf. multiline).

Virulence The observed infective capacity of individual races or strains of a pathogen when applied to suitable host tissues.

References

A'Brook, J. (1973). The effect of plant spacing on the number of aphids trapped over cocksfoot and kale crops. *Annals of Applied Biology*, **74**, 279–85.

Adams, E. B. & Line, R. F. (1984). Epidemiology and host morphology in the parasitism of rush skeletonweed by *Puccinia chondrillina*. *Phytopathology*, **74**, 745–8.

Agricultural Development and Advisory Service (1976). *Manual of Plant Growth Stage and Disease Assessment Keys*. Harpenden: ADAS.

Agrios, G. N. (1980). Escape from disease. In *Plant Disease: An Advanced Treatise*, vol. 5, ed. J. G. Horsfall & E. B. Cowling, pp. 17–37. New York: Academic Press.

Ahmad, I., Owera, S. A. P., Farrer, J. F. & Whitbread, R. (1982). The distribution of five major nutrients in barley plants infected with brown rust. *Physiological Plant Pathology*, **21**, 335–46.

Albershiem, P., Jones, T. M. & English, P. D. (1969). Biochemistry of the cell wall in relation to infective processes. *Annual Review of Phytopathology*, **7**, 171–94.

Alexander, H. M. (1984). Spatial patterns of disease induced by *Fusarium moniliforme* var. *subglutinans* in a population of *Plantago lanceolata*. *Oecologia* (*Berlin*), **62**, 141–3.

Alexander, H. M. & Burdon, J. J. (1984). The effect of disease induced by *Albugo candida* (white rust) and *Peronospora parasitica* (downy mildew) on the survival and reproduction of *Capsella bursa-pastoris* (shepherd's purse). *Oecologia* (*Berlin*), **64**, 314–18.

Alexander, H. M., Groth, J. V. & Roelfs, A. P. (1985). Virulence changes in *Uromyces appendiculatus* after five asexual generations on a partially resistant cultivar of *Phaseolus vulgaris*. *Phytopathology*, **75**, 449–53.

Allard, R. W., Babbel, G. R., Clegg, M. T. & Kahler, A. L. (1972). Evidence for coadaptation in *Avena barbata*. *Proceedings of the National Academy of Sciences, USA*, **69**, 3043–8.

Anderson, R. M. (1978). The regulation of host population growth by parasitic species. *Parasitology*, **76**, 119–57.

Anderson, R. M. (1979). Parasite pathogenicity and the depression of host population equilibra. *Nature*, **279**, 150–2.

Anderson, R. M. & May, R. M. (1978). Regulation and stability of host–parasite population interactions. I. Regulatory processes. *Journal of Animal Ecology*, **47**, 219–47.

Anderson, R. M. & May, R. M. (1979). Population biology of infectious diseases: Part I. *Nature*, **280**, 361–7.

Anderson, R. M. & May, R. M. (1981). The population dynamics of microparasites and their invertebrate hosts. *Philosophical Transactions of the Royal Society of London, Series B, Biological Sciences*, **291**, 451–524.

Anderson, R. M. & May, R. M. (1982). Coevolution of hosts and parasites. *Parasitology*, **85**, 411–26.

Antonovics, J. & Bradshaw, A. D. (1970). Evolution in closely adjacent plant populations. VIII. Clinal patterns at a mine boundary. *Heredity*, **25**, 349–62.

Atkinson, T. G. & Grant, M. N. (1967). An evaluation of streak mosaic losses in winter wheat. *Phytopathology*, **57**, 188–92.

Augspurger, C. K. (1983*a*). Seed dispersal of the tropical tree, *Platypodium elegans*, and the escape of its seedlings from fungal pathogens. *Journal of Ecology*, **71**, 759–71.

Augspurger, C. K. (1983*b*). Offspring recruitment around tropical trees: changes in cohort distance with time. *Oikos*, **40**, 189–96.

Augspurger, C. K. (1984). Seedling survival of tropical tree species: interactions of dispersal distance, light-gaps, and pathogens. *Ecology*, **65**, 1705–12.

Augspurger, C. K. & Kelly, C. K. (1984). Pathogen mortality of tropical tree seedlings: experimental studies of the effects of dispersal distance, seedling density, and light conditions. *Oecologia* (*Berlin*), **61**, 211–17.

Ayanru, D. K. G. & Browning, J. A. (1977). Effect of heterogeneous oat populations on the epiphytotic development of Victoria blight. *New Phytologist*, **79**, 613–23.

Ayres, P. G. (1978). Water relations of diseased plants. In *Water Deficits and Plant Growth*, vol. 5, ed. T. T. Kozlowski, pp. 1–60. London: Academic Press.

Ayres, P. G. (1981). Effects of disease on plant water relations. In *Effects of Disease on the Physiology of the Growing Plant*, ed. P. G. Ayres, pp. 131–48. Society for Experimental Biology Seminar Series, No. 11. Cambridge: Cambridge University Press.

Baker, H. G. (1947). Infection of species of *Melandrium* by *Ustilago violacea* (Pers.) Fuckel and the transmission of the resultant disease. *Annals of Botany N.S.*, **11**, 333–48.

Barber, J. C. (1966). *Variation among half-sib families from three loblolly pine stands in Georgia*. Georgia Forest Research Council Paper No. 37.

Barrett, J. A. (1980). Pathogen evolution in multilines and variety mixtures. *Journal of Plant Diseases and Protection*, **87**, 383–96.

Barrett, J. A. (1984). The genetics of host–parasite interaction. In *Evolutionary Ecology*, ed. B. Shorrocks, pp. 275–94. Oxford: Blackwell Scientific Publications.

Barrett, J. A. (in press). The gene-for-gene hypothesis: parable or paradigm. In *Ecology and Genetics of Host–Parasite Interactions*, ed. D. Rollinson.

Barrett, J. A. & Wolfe, M. S. (1978). Multilines and super-races – a reply. *Phytopathology*, **68**, 1535–7.

Barrett, J. A. & Wolfe, M. S. (1980). Pathogen response to host resistance and its implication in breeding programmes. *EPPO Bulletin*, **10**, 341–7.

Barton, L. V. (1961). *Seed Preservation and Longevity*. London: Leonard Hill.

Bateson, D. F. & Lumsden, R. D. (1965). Relation of calcium content and nature of pectic substances in bean hypocotyls of different ages to susceptibility to an isolate of *Rhizoctonia solani*. *Phytopathology*, **55**, 734–8.

Bawden, F. C. & Roberts, F. M. (1947). The influence of light intensity on the susceptibility of plants to certain viruses. *Annals of Applied Botany*, **34**, 286–96.

Ben-Kalio, V. D. & Clarke, D. D. (1979). Studies on tolerance in wild plants: effects of *Erysiphe fischeri* on the growth and development of *Senecio vulgaris*. *Physiological Plant Pathology*, **14**, 203–11.

Berg, L. A., Gough, F. J. & Williams, N. D. (1963). Inheritance of stem rust resistance in two wheat cultivars, Marquis and Kota. *Phytopathology*, **53**, 904–8.

Berger, R. D. (1973). Infection rates of *Cercospora apii* in mixed populations of susceptible and tolerant celery. *Phytopathology*, **63**, 535–7.

Bernard, R. L., Smith, P. E., Kaufmann, M. J. & Schmitthenner, A. F. (1957). Inheritance of resistance to *Phytophthora* root and stem rot in the soybean. *Agronomy Journal*, **49**, 391.

Bingham, R. T., Hoff, R. J. & McDonald, G. I. (1971). Disease resistance in forest trees. *Annual Review of Phytopathology*, **9**, 433–52.

Black, J. N. (1956). The influence of seed size and depth of sowing on pre-emergence and early vegetative growth of subterranean clover (*Trifolium subterraneum* L.). *Australian Journal of Agricultural Research*, **7**, 98–109.

Black, J. N. (1957). The early vegetative growth of three strains of subterranean clover (*Trifolium subterraneum* L.) in relation to size of seed. *Australian Journal of Agricultural Research*, **8**, 1–14.

Black, W., Mastenbroek, C., Mills, W. R. & Peterson, L. C. (1953). A proposal for an international nomenclature of races of *Phytophthora infestans* and of genes controlling immunity in *Solanum demissum* derivatives. *Euphytica*, **2**, 173–9.

Bloomberg, W. J. (1973). *Fusarium* root rot of Douglas-fir seedlings. *Phytopathology*, **63**, 337–41.

Boone, D. M. & Keitt, G. W. (1957). *Venturia inaequalis* (Cke). Wint. XII. Genes controlling pathogenicity of wild-type lines. *Phytopathology*, **47**, 403–9.

Bradshaw, A. D. (1959). Population differentiation in *Agrostis tenuis* Sibth. II. The incidence and significance of infection by *Epichloe typhina*. *New Phytologist*, **58**, 310–15.

Brinkerhoff, L. A. (1970). Variation in *Xanthomonas malvacearum* and its relation to control. *Annual Review of Phytopathology*, **8**, 85–110.

Brokenshire, T. (1974). Predisposition of wheat to *Septoria* infection following attack by *Erysiphe*. *Transactions of the British Mycological Society*, **63**, 393–7.

Bronnimann, A. (1968). Zur Kenntnis von *Septoria nodorum* Berk., dem Erreger der Spelzenbraune und einer Blattdurre des Weizens. *Phytopathologische Zeitschrift*, **61**, 101–46.

Brooks, D. H. (1972). Observations on the effects of mildew, *Erysiphe graminis*, on growth of spring and winter barley. *Annals of Applied Biology*, **70**, 149–56.

Brown, A. H. D., Feldman, M. W. & Nevo, E. (1980). Multilocus structure of natural populations of *Hordeum spontaneum*. *Genetics*, **96**, 523–36.

Brown, J. F. & Shipton, W. A. (1964). Relationship of penetration to infection type when seedling wheat leaves are inoculated with *Puccinia graminis tritici*. *Phytopathology*, **54**, 89–91.

Brown, V. K. (1982). The phytophagous insect community and its impact on early successional habitats. In *Proceedings of the Fifth International Symposium on Insect–Plant Relationships, Wageningen, 1982*, pp. 205–13. Wageningen: Pudoc.

Browning, J. A. (1974). Relevance of knowledge about natural ecosystems to development of pest management programs for agro-ecosystems. *Proceedings of the American Phytopathological Society*, **1**, 191–9.

Browning, J. A. (1979). Genetic protective mechanisms of plant-pathogen populations: their coevolution and use in breeding for resistance. In *Biology and Breeding for Resistance to Arthropods and Pathogens in Agricultural Plants*, ed. M. K. Harris, pp. 52–75. College Station, Texas: Texas A & M University.

Browning, J. A., Simons, M. D. & Torres, E. (1977). Managing host genes: epidemiologic and genetic concepts. In *Plant Disease: An Advanced Treatise*, vol. 1, ed. J. G. Horsfall & E. D. Cowling, pp. 191–212. New York: Academic Press.

Burdon, J. J. (1980). Variation in disease-resistance within a population of *Trifolium repens*. *Journal of Ecology*, **68**, 737–44.

Burdon, J. J. (1982). The effect of fungal pathogens on plant communities. In *The Plant Community as a Working Mechanism*, ed. E. I. Newman, pp. 99–112. Oxford: Blackwell Scientific Publications.

Burdon, J. J. (1985). Pathogens and the genetic structure of plant populations. In *Studies on Plant Demography: A Festschrift for John L. Harper*, ed. J. White, pp. 313–25. London: Academic Press.

Burdon, J. J. & Chilvers, G. A. (1975). Epidemiology of damping-off disease (*Pythium irregulare*) in relation to density of *Lepidium sativum* seedlings. *Annals of Applied Biology*, **81**, 135–43.

Burdon, J. J. & Chilvers, G. A. (1976*a*). Controlled environment experiments on epidemics of barley mildew in different density host stands. *Oecologia (Berlin)*, **26**, 61–72.

Burdon, J. J. & Chilvers, G. A. (1976*b*). Epidemiology of *Pythium*-induced damping-off in mixed species seedlings stands. *Annals of Applied Biology*, **82**, 233–40.

Burdon, J. J. & Chilvers, G. A. (1976*c*). The effect of clumped planting patterns on epidemics of damping-off disease in cress seedlings. *Oecologia (Berlin)*, **23**, 17–29.

Burdon, J. J. & Chilvers, G. A. (1977*a*). Controlled environment experiments on epidemic rates of barley mildew in different mixtures of barley and wheat. *Oecologia (Berlin)*, **28**, 141–6.

Burdon, J. J. & Chilvers, G. A. (1977*b*). The effect of barley mildew on barley and wheat competition in mixtures. *Australian Journal of Botany*, **25**, 59–65.

Burdon, J. J. & Chilvers, G. A. (1982). Host density as a factor in plant disease ecology. *Annual Review of Phytopathology*, **20**, 143–66.

Burdon, J. J., Groves, R. H. & Cullen, J. M. (1981). The impact of biological control on the distribution and abundance of *Chondrilla juncea* in south-eastern Australia. *Journal of Applied Ecology*, **18**, 957–66.

Burdon, J. J., Groves, R. H., Kaye, P. E. & Speer, S. S. (1984). Competition in mixtures of susceptible and resistant genotypes of *Chondrilla juncea* differentially infected with rust. *Oecologia (Berlin)*, **64**, 199–203.

Burdon, J. J., Luig, N. H. & Marshall, D. R. (1983). Isozyme uniformity and virulence variation in *Puccinia graminis* f.sp. *tritici* and *P. recondita* f.sp. *tritici* in Australia. *Australian Journal of Biological Sciences*, **36**, 403–10.

Burdon, J. J. & Marshall, D. R. (1981*a*). Inter- and intra-specific diversity in the disease-response of *Glycine* species to the leaf-rust fungus *Phakopsora pachyrhizi*. *Journal of Ecology*, **69**, 381–90.

Burdon, J. J. & Marshall, D. R. (1981*b*). Biological control and the reproductive mode of weeds. *Journal of Applied Ecology*, **18**, 649–58.

Burdon, J. J., Marshall, D. R. & Groves, R. H. (1980). Isozyme variation in *Chondrilla juncea* L. in Australia. *Australian Journal of Botany*, **28**, 193–8.

Burdon, J. J. & Muller, W. J. (in press). Measuring the cost of resistance to *Puccinia coronata* Cda in *Avena fatua* L. *Journal of Applied Ecology*.

Burdon, J. J., Oates, J. D. & Marshall, D. R. (1983). Interactions between *Avena* and *Puccinia* species. I. The wild hosts: *Avena barbata* Pott ex Link, *A. fatua* L. and *A. ludoviciana* Durieu. *Journal of Applied Ecology*, **20**, 571–84.

Burdon, J. J. & Shattock, R. C. (1980). Disease in plant communities. *Applied Biology*, **5**, 145–219.

Burdon, J. J. & Speer, S. S. (1984). A set of differential *Glycine* hosts for the identification of *Phakopsora pachyrhizi*. *Euphytica*, **33**, 891–6.

Burdon, J. J. & Whitbread, R. (1979). Rates of increase of barley mildew in mixed stands of barley and wheat. *Journal of Applied Ecology*, **16**, 253–8.

Butler, E. J. & Jones, S. G. (1955). *Plant Pathology*. London: Macmillan.

Caldwell, R. M., Schafer, J. F., Compton, L. E. & Patterson, F. L. (1958). Tolerance to cereal leaf rusts. *Science*, **128**, 714–5.

Cantlon, J. E. (1969). The stability of natural populations and their sensitivity to technology. *Brookhaven Symposia in Biology*, **22**, 197–205.

Carver, T. L. W. & Carr, A. J. H. (1978). Effects of host resistance on the development of haustoria and colonies of oat mildew. *Annals of Applied Biology*, **88**, 171–8.

Caten, C. E. (1974). Intra-racial variation in *Phytophthora infestans* and adaptation to field resistance for potato blight. *Annals of Applied Biology*, **77**, 259–70.

Catherall, P. L. (1966). Effects of barley yellow dwarf virus on the growth and yield of

single plants and simulated swards of perennial ryegrass. *Annals of Applied Biology*, **57**, 155–62.
Chamblee, D. S. (1958). Some above- and below-ground relationships of an Alfalfa–Orchardgrass mixture. *Agronomy Journal*, **50**, 434–7.
Chaplin, J. F. (1970). Associations among disease resistance, agronomic characteristics and chemical constituents in flue-cured tobacco. *Agronomy Journal*, **62**, 87–91.
Charlesworth, B. (1976). Recombination modification in a fluctuating environment. *Genetics*, **83**, 181–95.
Chi, C. C. & Hanson, E. W. (1962). Interrelated effects of environment and age of alfalfa and red clover seedlings on susceptibility to *Pythium debaryanum*. *Phytopathology*, **52**, 985–9.
Chilvers, G. A. & Brittain, E. G. (1972). Plant competition mediated by host-specific parasites – a simple model. *Australian Journal of Biological Sciences*, **25**, 749–56.
Chin, K. M. & Wolfe, M. S. (1984*a*). The spread of *Erysiphe graminis* f.sp. *hordei* in mixtures of barley varieties. *Plant Pathology*, **33**, 89–100.
Chin, K. M. & Wolfe, M. S. (1984*b*). Selection on *Erysiphe graminis hordei* in pure and mixed stands of barley. *Plant Pathology*, **33**, 535–46.
Christensen, C. M. (1978). Moisture and seed decay. In *Water Deficits and Plant Growth*, vol. 5, ed. T. T. Kozlowski, pp. 199–219. New York: Academic Press.
Clark, R. V. (1966). The influence of disease incidence and host tolerance on oat yields. *Canadian Plant Disease Survey*, **46**, 105–9.
Clark, R. V. (1980). Comparison of spot blotch severity in barley grown in pure stands and in mixtures with oats. *Canadian Journal of Plant Pathology*, **2**, 37–8.
Clarke, B. (1976). The ecological genetics of host–parasite relationships. In *Genetic Aspects of Host–Parasite Relationships*, ed. A. E. R. Taylor & R. Muller, pp. 87–103. Oxford: Blackwell Scientific Publications.
Clarke, D. D. & Harry, I. (1979). The distribution of resistance and virulence types in the *Senecio vulgaris/Erysiphe fischeri* host parasite system. In *Ninth International Congress of Plant Protection, Washington, D.C.*, August, 1979, abstract no. 242.
Clegg, M. T., Allard, R. W. & Kahler, A. L. (1972). Is the gene the unit of selection? Evidence from two experimental plant populations. *Proceedings of the National Academy of Sciences, USA*, **69**, 2474–8.
Clements, R. O., French, N., Guile, C. T., Golightly, W. H., Lewis, S. & Savage, M. J. (1982). The effect of pesticides on establishment of grass swards in England and Wales. *Annals of Applied Biology*, **101**, 305–13.
Clifford, B. C. (1968). Relations of disease resistance mechanisms to pathogen dynamics in oat crown rust epidemiology. *Dissertation Abstracts*, **298**, 835–6.
Clifford, B. C. (1972). The histology of race non-specific resistance to *Puccinia hordei* Otth. in barley. In *Proceedings of the European Mediterranean Cereal Rusts Conference*, Prague, 1972, vol. 1, 75–9.
Cobb, F. W. Jr, Slaughter, G. W., Rowney, D. L. & DeMars, C. J. (1982). Rate of spread of *Ceratocystis wageneri* in ponderosa pine stands in the central Sierra Nevada. *Phytopathology*, **72**, 1359–62.
Colhoun, J. (1973). Effects of environmental factors on plant disease. *Annual Review of Phytopathology*, **11**, 343–64.
Colhoun, J. (1979). Predisposition by the environment. In *Plant Disease: An Advanced Treatise*, vol. 4, ed. J. G. Horsfall & E. B. Cowling, pp. 75–96. New York: Academic Press.
Commonwealth Mycological Institute (1968). *Plant Pathologist's Pocketbook*. Kew: Commonwealth Mycological Institute.

Conard, S. G. & Radosevich, S. R. (1979). Ecological fitness of *Senecio vulgaris* and *Amaranthus retroflexus* biotypes susceptible or resistant to atrazine. *Journal of Applied Ecology*, **16**, 171–7.

Cook, R. J. & Baker, K. F. (1983). *The Nature and Practice of Biological Control of Plant Pathogens*. St Paul, Minnesota: American Phytopathological Society.

Cournoyer, B. M. (1970). Crown rust epiphytology with emphasis on the quantity and periodicity of spore dispersal from heterogeneous oat cultivar – rust race populations. Ph.D. thesis, Iowa State University.

Coyne, D. P., Steadman, J. R. & Schwartz, H. F. (1978). Effect of genetic blends of dry beans (*Phaseolus vulgaris*) of different plant architecture on apothecia production of *Sclerotinia sclerotiorum* and white mold infection. *Euphytica*, **27**, 225–31.

Crawley, M. J. (1983). *Herbivory: The Dynamics of Animal and Plant Interactions*. Oxford. Blackwell Scientific Publications.

Crill, P. (1977). An assessment of stabilizing selection in crop variety development. *Annual Review of Phytopathology*, **15**, 185–202.

Crute, I. R. & Johnson, A. G. (1976). The genetic relationship between races of *Bremia lactucae* and cultivars of *Lactuca sativa*. *Annals of Applied Biology*, **83**, 125–37.

Cullen, J. M. & Groves, R. H. (1977). The population biology of *Chondrilla juncea* L. in Australia. *Proceedings of the Ecological Society of Australia*, **10**, 121–34.

Czabator, F. J. (1971). Fusiform rust of southern pines – a critical review. *United States Department of Agriculture Forest Service Research Paper* SO-65.

Daly, J. M. (1976). The carbon balance of diseased plants: changes in respiration, photosynthesis and translocation. In *Encyclopedia of Plant Physiology*, vol. 4, ed. R. Heitefuss & P. H. Williams, pp. 450–79. Berlin: Springer-Verlag.

Day, F. P., Jr & Monk, C. D. (1974). Vegetation patterns on a southern Appalachian watershed. *Ecology*, **55**, 1064–74.

Day, K. L. (1981). Spring barley variety mixtures as a means of powdery mildew control. *Journal of National Institute of Agricultural Botany*, **15**, 421–9.

Day, P. R. (1956). Race names of *Cladosporium fulvum*. *Tomato Genetics Co-operative Report*, **6**, 13–14.

Day, P. R. (1960). Variation in phytopathogenic fungi. *Annual Review of Microbiology*, **14**, 1–16.

Day, P. R. (1974). *Genetics of Host–Parasite Interaction*. San Francisco: W. H. Freeman.

Day, P. R. (1978). The genetic basis of epidemics. In *Plant Disease: An Advanced Treatise*, vol. 2, ed. J. G. Horsfall & E. B. Cowling, pp. 263–85, New York: Academic Press.

Day, P. R., Barrett, J. A. & Wolfe, M. S. (1983). The evolution of host–parasite interaction. In *Genetic Engineering in Plants*, ed. T. Kosuge, C. P. Meredith & A. Hollaender, pp. 419–30. New York: Plenum Press.

de Cubillos, F. C. & Thurston, H. D. (1975). The effects of viruses on infection by *Phytophthora infestans* (Mont.) de Bary in potatoes. *American Potato Journal*, **52**, 221–6.

de Wit, C. T. (1960). *On Competition*. Wageningen: Pudoc.

Deevey, E. S. (1947). Life tables for natural populations of animals. *Quarterly Review of Biology*, **22**, 283–314.

Dinoor, A. (1970). Sources of oat crown rust resistance in hexaploid and tetraploid wild oats in Israel. *Canadian Journal of Botany*, **48**, 153–61.

Dinoor, A. (1977). Oat crown rust resistance in Israel. *Annals of the New York Academy of Sciences*, **287**, 357–66.

Dinoor, A. & Eshed, N. (1984). The role and importance of pathogens in natural plant communities. *Annual Review of Phytopathology*, **22**, 443–66.

Dinus, R. J. (1974). Knowledge about natural ecosystems as a guide to disease control in managed forests. *Proceedings of the American Phytopathological Society*, **1**, 184–90.

Doling, D. A. (1964). The influence of seedling competition on the amount of loose smut (*Ustilago nuda* (Jens.) Rostr.) appearing in barley crops. *Annals of Applied Biology*, **54**, 91–8.

Dow, R. L., Porter, D. M. & Powell, N. L. (1981). Effect of thinning on *Sclerotinia* blight of peanut. *Phytopathology*, **71**, 766.

Dowling, P. N. & Linscott, D. L. (1983). Use of pesticides to determine relative importance of pest and disease factors limiting establishment of sod-seeded lucerne. *Grass and Forage Science*, **38**, 179–85.

Duff, A. D. S. (1954). Seedling resistance and mature-plant susceptibility of wheat to *Puccinia graminis* found in Kenya. *Nature*, **173**, 779.

Dukes, P. D. & Apple, J. L. (1961). Chemotaxis of zoospores of *Phytophthora parasitica* var. *nicotianae* by plant roots and certain chemical solutions. *Phytopathology*, **51**, 195–7.

Ellingboe, A. H. (1976). Genetics of host–parasite interactions. In *Encyclopedia of Plant Physiology*, vol. 4, ed. R. Heitefuss & P. H. Williams, pp. 761–78. Berlin: Springer-Verlag.

Falloon, R. E. (1976). Effect of infection by *Ustilago bullata* on vegetative growth of *Bromus catharticus*. *New Zealand Journal of Agricultural Research*, **19**, 249–54.

Federation of British Plant Pathologists (1973). *A Guide to the Use of Terms in Plant Pathology*. Phytopathological Papers No. 17. Kew: Commonwealth Mycological Institute.

Felsenstein, J. & Yokoyama, S. (1976). The evolutionary advantage of recombination. II. Individual selection for recombination. *Genetics*, **83**, 845–59.

Fischbeck, G., Schwarzbach, E., Sobel, Z. & Wahl, I. (1976). Types of protection against barley powdery mildew in Germany and Israel selected from *Hordeum spontaneum*. In *Proceedings, Third International Barley Genetics Symposium, Barley Genetics*, Garching 1975, pp. 412–17. Munich: Verlag Karl Thiemig.

Flanges, A. L. & Dickson, J. G. (1961). The genetic control of pathogenicity, serotypes and variability in *Puccinia sorghi*. *American Journal of Botany*, **48**, 275–85.

Fleming, R. A. (1980). Selection pressures and plant pathogens: robustness of the model. *Phytopathology*, **70**, 175–8.

Fleming, R. A., Marsh, L. M. & Tuckwell, H. C. (1982). Effect of field geometry on the spread of crop disease. *Protection Ecology*, **4**, 81–108.

Flor, H. H. (1942). Inheritance of pathogenicity in *Melampsora lini*. *Phytopathology*, **32**, 653–69.

Flor, H. H. (1951). Genes for resistance to rust in Victory flax. *Agronomy Journal*, **43**, 527–31.

Flor, H. H. (1956). The complementary genic systems in flax and flax rust. *Advances in Genetics*, **8**, 29–54.

Flor, H. H. (1971). Current status of the gene-for-gene concept. *Annual Review of Phytopathology*, **9**, 275–96.

Foster, J. (1964). Studies on the population dynamics of the daisy (*Bellis perennis* L.). Ph.D. thesis, University of Wales.

Fric, F. (1975). Translocation of ^{14}C-labelled assimilates in barley plants infected with powdery mildew (*Erysiphe graminis* f.sp. *hordei* Marchal). *Phytopathologie Zeitschrift*, **84**, 88–95.

Fried, P. M., MacKenzie, D. R. & Nelson, R. R. (1981). Yield loss caused by *Erysiphe graminis* f. sp. *tritici* on single culms of 'Chancellor' wheat and four multilines. *Journal of Plant Diseases and Protection*, **88**, 256–64.

Gates, D. J., Westcott, M., Burdon, J. J. & Alexander, H. M. (1986). Competition and stability in plant mixtures in the presence of disease. *Oecologia* (*Berlin*), **68**, 559–66.

Gaumann, E. (1950). *Principles of Plant Infection*. London: Crosby Lockwood.

Gaunt, R. E. (1981). Disease tolerance – an indicator of thresholds? *Phytopathology*, **71**, 915–16.

Geiger, R. (1965). *The Climate Near the Ground.* Cambridge, Massachusetts: Harvard University Press.

Gheorghies, C. (1972). [Research concerning the influence of certain soil and crop factors upon the *Septoria* leaf blotch of wheat.] *Lucrari Stientifice*, **15**, 113–19. (In Romanian.)

Gibbs, A. (1980). A plant virus that partially protects its wild legume host against herbivores. *Intervirology*, **13**, 42–7.

Gibbs, A. (1983). Virus ecology – 'struggle' of the genes. In *Encyclopedia of Plant Physiology*, new series, vol. 12C, ed. O. L. Lange, P. S. Nobel, C. B. Osmond & H. Ziegler, pp. 537–58. Berlin: Springer-Verlag.

Gibbs, A. & Harrison, B. (1976). *Plant Virology: The Principles.* London: Edward Arnold.

Gibson, I. A. S. (1956). Sowing density and damping-off in pine seedlings. *East African Agricultural Journal*, **21**, 183–8.

Gill, C. C. & Westdal, P. H. (1966). Effect of temperature on symptom expression of barley infected with aster yellows or barley yellow dwarf viruses. *Phytopathology*, **56**, 369–70.

Graham, R. D. (1983). Effects of nutrient stress on susceptibility of plants to disease with particular reference to the trace elements. *Advances in Botanical Research*, **10**, 221–76.

Gram, E. (1960). Quarantines. In *Plant Pathology: An Advanced Treatise*, vol. 3, ed. J. G. Horsfall & A. E. Dimond, pp. 313–56. New York: Academic Press.

Grant, M. W. & Archer, S. A. (1983). Calculation of selection coefficients against unnecessary genes for virulence from field data. *Phytopathology*, **73**, 547–51.

Green, G. J. (1979). Stem rust of wheat, barley and rye in Canada in 1978. *Canadian Plant Disease Survey*, **59**, 43–7.

Gregory, P. H. (1968). Interpreting plant disease gradients. *Annual Review of Phytopathology*, **6**, 189–212.

Greig-Smith, P. (1964). *Quantitative Plant Ecology*, 2nd edn. London: Butterworths.

Groenewegen, L. J. M. & Zadoks, J. C. (1979). Exploiting within-field diversity as a defense against cereal diseases: a plea for "poly-genotype" varieties. *Indian Journal of Genetics and Plant Breeding*, **39**, 81–94.

Groth, J. V. (1976). Multilines and "super races": a simple model. *Phytopathology*, **66**, 937–9.

Groth, J. V. & Person, C. O. (1977). Genetic interdependence of host and parasite in epidemics. *Annals of the New York Academy of Sciences*, **287**, 97–106.

Groth, J. V. & Roelfs, A. P. (1982). The effect of sexual and asexual reproduction on race abundance in cereal rust fungus populations. *Phytopathology*, **72**, 1503–7.

Groves, R. H. & Cullen, J. M. (1981). *Chondrilla juncea*: the ecological control of a weed. In *The Ecology of Pests*, ed. R. L. Kitching & R. E. Jones, pp. 7–17. Melbourne: CSIRO.

Groves, R. H. & Williams, J. D. (1975). Growth of skeleton weed (*Chondrilla juncea* L.) as affected by growth of subterranean clover (*Trifolium subterraneum* L.) and infection by *Puccinia chondrillina* Bubak & Syd. *Australian Journal of Agricultural Research*, **26**, 975–83.

Grummer, G. & Roy, S. K. (1966). Intervarietal mixtures of rice and incidence of brown-spot disease (*Helminthosporium oryzae* Breda de Haan). *Nature*, **209**, 1265–7.

Gustafsson, M. & Larsson, C. (1984). Variation in the patterns of virulence and the relative fitness of virulence phenotypes in Swedish populations of *Bremia lactucae*. *Hereditas*, **101**, 9–17.

Hamblin, J., Wood, P. McR. & Allen, J. G. (1981). The use of interspecific mixtures of *Lupinus* species to simulate the effects of different levels of *Phomopsis leptostromiformis* resistance in *L. angustifolius*. *Euphytica*, **30**, 203–7.

Hamilton, W. D. (1980). Sex versus non-sex versus parasite. *Oikos*, **35**, 282–90.

Hamilton, W. D. (1982). Pathogens as causes of genetic diversity in their host populations. In *Population Biology of Infectious Diseases*, ed. R. M. Anderson & R. M. May, pp. 269–96. Dahlem Konferenzen 1982. Berlin: Springer-Verlag.
Hampson, M. C. (1980). Pathogenesis of *Synchytrium endobioticum*. 2. Effect of soil amendments and fertilization. *Canadian Journal of Plant Pathology*, **2**, 148–51.
Hansen, L. R. & Magnus, H. A. (1973). Virulence spectrum of *Rhynchosporium secalis* in Norway and sources of resistance in barley. *Phytopathologische Zeitschrift*, **76**, 303–13.
Harlan, H. V. & Martini, M. I. (1938). The effect of natural selection in a mixture of barley varieties. *Journal of Agricultural Research*, **57**, 189–99.
Harlan, J. R. (1976). Diseases as a factor in plant evolution. *Annual Review of Phytopathology*, **14**, 31–51.
Harper, J. L. (1977). *Population Biology of Plants*. London: Academic Press.
Harrison, J. G. (1977). The effect of seed deterioration on the growth of barley. *Annals of Applied Biology*, **87**, 485–94.
Hartley, C. P. (1921). Damping-off in forest nurseries. *United States Department of Agriculture. Bulletin* No. 934.
Hassan, A. & MacDonald, J. A. (1971). *Ustilago violacea* on *Silene dioica*. *Transactions of the British Mycological Society*, **56**, 451–61.
Hassell, M. P. (1980). Some consequences of habitat heterogeneity for population dynamics. *Oikos*, **35**, 150–60.
Hastings, A. (1977). Spatial heterogeneity and the stability of predator–prey systems. *Theoretical Population Biology*, **12**, 37–48.
Hayden, E. B., Jr (1956). Progressive development of infection by *Puccinia graminis* var. *tritici* Ericks. and E. Henn. (Guyot) on certain varieties of wheat and the relation of stem rust to yield. *Dissertation Abstracts*, **16**, 2262–3.
Heringa, R. J., Van Norel, A. & Tazelaar, M. F. (1969). Resistance to powdery mildew. (*Erysiphe polygoni* D.C.) in peas (*Pisum sativum* L.). *Euphytica*, **18**, 163–9.
Hickey, D. A. & McNeilly, T. (1975). Competition between metal tolerant and normal plant populations; a field experiment on normal soil. *Evolution*, **29**, 458–64.
Hickman, J. C. (1979). The basic biology of plant numbers. In *Topics in Plant Population Biology*, ed. O. T. Solbrig, S. Jain, G. B. Johnson & P. H. Raven, pp. 232–63. New York: Columbia University Press.
Hilborn, R. (1975). The effect of spatial heterogeneity on the persistence of predator–prey interactions. *Theoretical Population Biology*, **8**, 346–55.
Hirst, J. M., Hide, G. A., Stedman, O. J. & Griffith, R. L. (1973). Yield compensation in gappy potato crops and methods to measure effects of fungi pathogenic on seed tubers. *Annals of Applied Biology*, **73**, 143–50.
Holton, C. S. & Halisky, P. M. (1960). Dominance of avirulence and monogenic control of virulence in race hybrids of *Ustilago avenae*. *Phytopathology*, **50**, 766–70.
Holton, C. S., Hoffman, J. A. & Duran, R. (1968). Variation in the smut fungi. *Annual Review of Phytopathology*, **6**, 213–42.
Hooker, A. L. (1967). The genetics and expression of resistance in plants to rusts of the genus *Puccinia*. *Annual Review of Phytopathology*, **5**, 163–82.
Hooker, W. J. & Fronek, F. R. (1961). The influence of virus Y infection on early blight susceptibility in potato. In *Proceedings of the Fourth Potato Virus Diseases Conference*, pp. 76–81.
Howard, H. W. (1968). The relation between resistance genes in potatoes and pathotypes of potato-root eelworm (*Heterodera rostochiensis*), wart disease (*Synchytrium endobioticum*) and potato virus X. In *Abstracts of the First International Congress of Plant Pathology*, London, no. 92.
Huang, H. C. & Hoes, J. A. (1980). Importance of plant spacing and sclerotial position to development of *Sclerotinia* wilt of sunflowers. *Plant Disease*, **64**, 81–4.

Huber, D. M. (1978). Disturbed mineral nutrition. In *Plant Disease: An Advanced Treatise*, vol. 3, ed. J. G. Horsfall & E. B. Cowling, pp. 163–81. New York: Academic Press.

Huber, D. M. & Watson, R. D. (1974). Nitrogen form and plant disease. *Annual Review of Phytopathology*, **12**, 139–65.

Hunt, R. S. & Van Sickle, G. A. (1984). Variation in susceptibility to sweet fern rust among *Pinus contorta* and *P. banksiana*. *Canadian Journal of Forestry Research*, **14**, 672–5.

Inouye, R. S. (1981). Interactions among unrelated species: granivorous rodents, a parasitic fungus and a shared prey species. *Oecologia (Berlin)*, **49**, 425–7.

Jackson, L. F. (1979). Studies on the nature of host resistance to barley scald and its influence on pathogenic complexity in *Rhynchosporium secalis*. Ph.D. thesis. University of California, Davis.

Jackson, L. F., Kahler, A. L., Webster, R. K. & Allard, R. W. (1978). Conservation of scald resistance in barley composite cross populations. *Phytopathology*, **68**, 645–50.

Jaenike, J. (1978). An hypothesis to account for the maintenance of sex within populations. *Evolutionary Theory*, **3**, 191–4.

James, W. C. (1971). An illustrated series of assessment keys for plant diseases, their preparation and usage. *Canadian Plant Disease Survey*, **51**, 39–65.

James, W. C. (1974). Assessment of plant diseases and losses. *Annual Review of Phytopathology*, **12**, 27–48.

James, W. C. & Teng, P. S. (1979). The quantification of production constraints associated with plant disease. *Applied Biology*, **4**, 201–67.

Janzen, D. H. (1968). Host plants as islands in evolutionary and contemporary time. *American Naturalist*, **102**, 592–5.

Janzen, D. H. (1970). Herbivores and the number of tree species in tropical forests. *American Naturalist*, **104**, 501–28.

Janzen, D. H. (1971). Seed predation by animals. *Annual Review of Ecology and Systematics*, **2**, 465–92.

Jarosz, A. M. (1984). Ecological and evolutionary dynamics of *Phlox – Erysiphe cichoracearum* interactions. Ph.D. thesis, Purdue University.

Jayakar, S. D. (1970). A mathematical model for interaction of gene frequencies in a parasite and its host. *Theoretical Population Biology*, **1**, 140–64.

Jeger, M. J., Jones, G. D. & Griffiths, E. (1981). Disease progress of non-specialised fungal pathogens in intraspecific mixed stands of cereal cultivars. II. Field experiments. *Annals of Applied Biology*, **98**, 199–210.

Jenkyn, J. F. (1970). Epidemiology of cereal powdery mildew (*Erysiphe graminis*). *Rothamsted Experimental Station Report*, 1969, Part 1, p. 151.

Jenkyn, J. F. & Bainbridge, A. (1978). Biology and pathology of cereal powdery mildews. In *The Powdery Mildews*, ed. D. M. Spencer, pp. 283–321. London: Academic Press.

Jennersten, O., Nilsson, S. G. & Wastljung, U. (1983). Local plant populations as ecological islands: the infection of *Viscaria vulgaris* by the fungus *Ustilago violacea*. *Oikos*, **41**, 391–5.

Johnson, R. (1978). Induced resistance to fungal diseases with special reference to yellow rust of wheat. *Annals of Applied Biology*, **89**, 107–10.

Johnson, R. & Allen, D. J. (1975). Induced resistance to rust diseases and its possible role in the resistance of multiline varieties. *Annals of Applied Biology*, **80**, 359–63.

Johnson, R. & Taylor, A. J. (1972). Isolates of *Puccinia striiformis* collected in England from the wheat varieties Maris Beacon and Joss Cambier. *Nature*, **238**, 105–6.

Johnson, R. & Taylor, A. J. (1976). Spore yield of pathogens in investigations of the race-specificity of host resistance. *Annual Review of Phytopathology*, **14**, 97–119.

Jones, D. G. & Jenkins, P. D. (1978). Predisposing effects of eyespot (*Pseudocercosporella herpotrichoides*) on *Septoria nodorum* infection of winter wheat. *Annals of Applied Biology*, **90**, 45–9.

Jones, I. T., O'Reilly, A. & Davies, I. J. E. R. (1983). Inheritance of adult plant resistance to oat mildew. *Welsh Plant Breeding Station Annual Report*, 1982, 103–5.

Jones, M. G. (1933). Grassland management and its influence on the sward. *Journal of the Royal Agricultural Society of England*, **94**, 24–41.

Kammeraad, J. W. & Brewer, R. (1963). Dispersal rate and elm density as factors in the occurrence of Dutch elm disease. *American Midland Naturalist*, **70**, 159–63.

Kassanis, B. (1952). Some effects of high temperature on the susceptibility of plants to infection with viruses. *Annals of Applied Biology*, **39**, 358–69.

Keane, E. M. & Sackston, W. E. (1970). Effects of boron and calcium nutrition of flax on *Fusarium* wilt. *Canadian Journal of Plant Science*, **50**, 415–22.

Keeling, B. L. (1974). Soybean seed rot and the relation of seed exudate to host susceptibility. *Phytopathology*, **64**, 1445–7.

Kermack, W. O. & McKendrick, A. G. (1927). A contribution to the mathematical theory of epidemics. *Proceedings of the Royal Society of London, Series A, Mathematical Sciences*, **115**, 700–21.

Kershaw, K. A. (1964). *Quantitative and Dynamic Ecology*. London: Edward Arnold.

King, T. J. (1977). The plant ecology of ant-hills in calcareous grasslands. III. Factors affecting the population sizes of selected species. *Journal of Ecology*, **65**, 279–315.

Kinloch, B. B. (1982). Mechanisms and inheritance of rust resistance in conifers. In *Resistance to Diseases and Pests in Forest Trees*, ed. H. M. Heybroek, B. R. Stephan & K. von Weissenberg, pp. 119–29. Wageningen: Pudoc.

Kinloch B. B. Jr & Stonecypher, R. W. (1969). Genetic variation in susceptibility to fusiform rust in seedlings from a wild population of loblolly pine. *Phytopathology*, **59**, 1246–55.

Kiyosawa, S. (1980). The possible application of gene-for-gene concept in blast resistance. *Japanese Agricultural Research Quarterly*, **14**, 9–14.

Klages, K. H. W. (1936). Changes in the proportions of the components of seeded and harvested cereal mixtures in abnormal seasons. *Agronomy Journal*, **28**, 935–40.

Kochman, J. K. & Brown, J. F. (1976*a*). Effect of temperature, light and host on prepenetration development of *Puccinia graminis avenae* and *Puccinia coronata avenae*. *Annals of Applied Biology*, **82**, 241–9.

Kochman, J. K. & Brown, J. F. (1976*b*). Host and environmental effects on the penetration of oats by *Puccinia graminis avenae* and *Puccinia coronata avenae*. *Annals of Applied Biology*, **82**, 251–8.

Kosuge, T. (1978). The capture and use of energy by diseased plants. In *Plant Disease: An Advanced Treatise*, vol. 3, ed. J. G. Horsfall & E. B. Cowling, pp. 85–116. New York: Academic Press.

Kranz, J. (1974). Comparison of epidemics. *Annual Review of Phytopathology*, **12**, 355–74.

Kreitlow, K. W., Garber, R. J. & Robinson, R. R. (1950). Investigations on seed treatment of alfalfa, red clover and Sudan grass for control of damping-off. *Phytopathology*, **40**, 883–98.

Kuc, J., Shockley, G. & Kearney, K. (1975). Protection of cucumber against *Colletotrichum lagenarium* by *Colletotrichum lagenarium*. *Physiological Plant Pathology*, **7**, 195–9.

Large, E. C. (1945). Field trials of copper fungicides for the control of potato blight. I. Foliage protection and yield. *Annals of Applied Biology*, **32**, 319–29.

Latch, G. C. M. & Potter, L. R. (1977). Interaction between crown rust (*Puccinia coronata*) and two viruses of ryegrass. *Annals of Applied Biology*, **87**, 139–45.

Laviolette, F. A. & Athow, K. L. (1971). Relationship of age of soybean seedlings and inoculum to infection by *Pythium ultimum*. *Phytopathology*, **61**, 439–40.

Lawrence, G. J., Mayo, G. M. E. & Shepherd, K. W. (1981). Interactions between genes controlling pathogenicity in the flax rust fungus. *Phytopathology*, **71**, 12–19.

Lawrence, W. H. & Rediske, J. H. (1962). Fate of sown Douglas-fir seed. *Forest Science*, **8**, 210–18.

Leach, L. D. (1947). Growth rates of host and pathogen as factors determining the severity of preemergence damping-off. *Journal of Agricultural Research*, **75**, 161–79.

Lee, J. A. (1981). Variation in the infection of *Silene dioica* (L.) Clairv. by *Ustilago violacea* (Pers.) Fuckel in north west England. *New Phytologist*, **87**, 81–9.

Leonard, K. J. (1969*a*). Factors affecting rates of stem rust increase in mixed plantings of susceptible and resistant oat varieties. *Phytopathology*, **59**, 1845–50.

Leonard, K. J. (1969*b*). Selection in heterogeneous populations of *Puccinia graminis* f. sp. *avenae*. *Phytopathology*, **59**, 1851–7.

Leonard, K. J. (1977). Selection pressures and plant pathogens. *Annals of the New York Academy of Sciences*, **287**, 207–22.

Leonard, K. J. (1984). Population genetics of gene-for-gene interactions between plant host resistance and pathogen virulence. In *Proceedings of the XV International Congress of Genetics*, New Delhi, December 1983, pp. 131–48. New Delhi: Oxford and IBH.

Leonard, K. J. (in press). The host population as a selective factor (including stabilizing selection). In *Populations of Plant Pathogens: Their Dynamics and Genetics*, ed. M. S. Wolfe & C. Caten. Oxford: Blackwell Scientific Publications.

Leonard, K. J. & Czochor, R. J. (1978). In response to 'Selection pressures and plant pathogens: stability of equilibria'. *Phytopathology*, **68**, 971–3.

Leonard, K. J. & Czochor, R. J. (1980). Theory of genetic interactions among populations of plants and their pathogens. *Annual Review of Phytopathology*, **18**, 237–58.

Levin, D. A. (1975). Pest pressure and recombination systems in plants. *American Naturalist*, **109**, 437–51.

Levin, S. A. (1983). Some approaches to the modelling of coevolutionary interactions. In *Coevolution*, ed. M. H. Nitecki, pp. 21–65. Chicago: University of Chicago Press.

Livne, A. & Daly, J. M. (1966). Translocation in healthy and rust-affected beans. *Phytopathology*, **56**, 170–5.

Luig, N. H. (1979). Rust survey 1977–78. *New South Wales Department of Agriculture, Biology Branch Disease Survey*, 1977–78, 10–17.

Luig, N. H. (1983). A survey of virulence genes in wheat stem rust, *Puccinia graminis* f. sp. *tritici*. *Advances in Plant Breeding*, **11**, 1–198.

Luig, N. H. & Rajaram, S. (1972). The effect of temperature and genetic background on host gene expression and interaction to *Puccinia graminis tritici*. *Phytopathology*, **62**, 1171–4.

Luig, N. H. & Watson, I. A. (1961). A study of inheritance of pathogenicity in *Puccinia graminis* var. *tritici*. *Proceedings of the Linnean Society of New South Wales*, **86**, 217–29.

Luig, N. H. & Watson, I. A. (1970). The effect of complex genetic resistance in wheat on the variability of *Puccinia graminis* f. sp. *tritici*. *Proceedings of the Linnean Society of New South Wales*, **95**, 22–45.

Luig, N. H. & Watson, I. A. (1977). The role of barley, rye and grasses in the 1973–74 wheat stem rust epiphytotic in southern and eastern Australia. *Proceedings of the Linnean Society of New South Wales*, **101**, 65–76.

Lukens, R. J. & Mullany, R. (1972). The influence of shade and wet soil on southern corn leaf blight. *Plant Disease Reporter*, **56**, 203–6.

Lupton, F. G. H. & Johnson, R. (1970). Breeding for mature-plant resistance to yellow rust in wheat. *Annals of Applied Biology*, **66**, 137–43.

Lupton, F. G. H. & Macer, R. C. F. (1962). Inheritance of resistance to yellow rust (*Puccinia glumarum* Erikss. & Henn.) in seven varieties of wheat. *Transactions of the British Mycological Society*, **45**, 21–45.

Lynch, J. M. (1978). Microbial interactions around imbibed seeds. *Annals of Applied Biology*, **89**, 165–7.

MacArthur, R. H. & Wilson, E. O. (1967). *The Theory of Island Biogeography*. Princeton: Princeton University Press.

Mack, R. N. & Pyke, D. A. (1984). The demography of *Bromus tectorum*: the role of microclimate, grazing and disease. *Journal of Ecology*, **72**, 731–48.

McNeilly, T. (1968). Evolution in closely adjacent plant populations. III. *Agrostis tenuis* on a small copper mine. *Heredity*, **23**, 99–108.

Malm, N. R. & Hooker, A. L. (1962). Resistance to rust, *Puccinia sorghi* Schw., conditioned by recessive genes in two corn inbred lines. *Crop Science*, **2**, 145–7.

Mannson, T. (1955). [Powdery mildew, *Erysiphe graminis* DC., on wheat.] *Sveriges Utsadesforenings Tidskrift*, **65**, 220–41. (In Swedish.)

Marshall, D. R. & Burdon, J. J. (1981). Multiline varieties and disease control. IV. Effects of reproduction of pathogen biotypes on resistant hosts on the evolution of virulence. *Sabrao Journal*, **13**, 116–26.

Marshall, D. R., Burdon, J. J. & Muller, W. J. (1986). Multiline varieties and disease control. VI. Effects of selection at different stages of the pathogen life cycle on the evolution of virulence. *Theoretical and Applied Genetics*, **71**, 801–9.

Marshall, D. R. & Pryor, A. J. (1978). Multiline varieties and disease control. I. The "dirty crop" approach with each component carrying a unique single resistance gene. *Theoretical and Applied Genetics*, **51**, 177–84.

Marshall, D. R. & Pryor A. J. (1979). Multiline varieties and disease control. II. The "dirty crop" approach with components carrying two or more genes for resistance. *Euphytica*, **28**, 145–59.

Martens, J. W., McKenzie, R. I. H. & Green, G. J. (1970). Gene-for-gene relationships in the *Avena*: *Puccinia graminis* host–parasite system in Canada. *Canadian Journal of Botany*, **48**, 969–75.

Mathre, D. E. (1978). Disrupted reproduction. In *Plant Disease: An Advanced Treatise*, vol. 3, ed. J. G. Horsfall & E. B. Cowling, pp. 257–78. New York: Academic Press.

Mathur, S. G. & Hansing, E. D. (1962). Effect of *Ustilago tritici* on the growth and morphological characteristics of winter wheat. *Phytopathology*, **52**, 20.

Matta, A. (1980). Defences triggered by previous diverse invaders. In *Plant Disease: An Advanced Treatise*, vol. 5, ed. J. G. Horsfall & E. B. Cowling, pp. 345–61. New York: Academic Press.

Matthews, R. E. F. (1970). *Plant Virology*. New York: Academic Press.

May, R. M. (1982). Introduction. In *Population Biology of Infectious Diseases*, ed. R. M. Anderson & R. M. May, pp. 1–12. Dahlem Konferenzen 1982. Berlin: Springer-Verlag.

May, R. M. & Anderson, R. M. (1978). Regulation and stability of host–parasite population interactions. II. Destabilizing processes. *Journal of Animal Ecology*, **47**, 249–67.

May, R. M. & Anderson, R. M. (1979). Population biology of infectious diseases: Part II. *Nature*, **280**, 455–61.

May, R. M. & Anderson, R. M. (1983). Epidemiology and genetics in the coevolution of parasites and hosts. *Proceedings of the Royal Society of London, Series B, Biological Sciences*, **219**, 281–313.

Maynard Smith, J. (1971). What use is sex? *Journal of Theoretical Biology*, **30**, 319–35.

Maynard Smith, J. (1978). *The Evolution of Sex*. Cambridge: Cambridge University Press.
Metzger, R. J. & Trione, E. J. (1962). Application of the gene-for-gene relationship hypothesis to the *Triticum–Tilletia* system. *Phytopathology*, **52**, 363.
Michail, S. H. & Carr, A. J. H. (1966). Effect of seed treatment on establishment of grass seedlings. *Plant Pathology*, **15**, 60–4.
Michel, L. J. & Simons, M. D. (1971). Relative tolerance of contemporary oat cultivars to currently prevalent races of crown rust. *Crop Science*, **11**, 99–100.
Miles, J. W. & Lenné, J. M. (1984). Genetic variation within a natural *Stylosanthes guianensis*, *Colletotrichum gloeosporioides* host–pathogen population. *Australian Journal of Agricultural Research*, **35**, 211–18.
Miller, P. (1966). The effect of weather on prevalence of disease. *The American Biology Teacher*, **28**, 469–72.
Mitchell, D. T. & Rice, K. A. (1979). Translocation of ^{14}C-labelled assimilates in cabbage during club root development. *Annals of Applied Biology*, **92**, 143–52.
Mode, C. J. (1958). A mathematical model for the co-evolution of obligate parasites and their hosts. *Evolution*, **12**, 158–65.
Mode, C. J. (1961). A generalized model of a host–pathogen system. *Biometrics*, **17**, 386–404.
Moseman, J. G. (1959). Host–pathogen interaction of the genes for resistance in *Hordeum vulgare* and for pathogenicity in *Erysiphe graminis* f.sp. *hordei*. *Phytopathology*, **49**, 469–72.
Moseman, J. G., Nevo, E., El Morshidy, M. A. & Zohary, D. (1984). Resistance of *Triticum dicoccoides* to infection with *Erysiphe graminis tritici*. *Euphytica*, **33**, 41–7.
Moseman, J. G., Nevo, E. & Zohary, D. (1983). Resistance of *Hordeum spontaneum* collected in Israel to infection with *Erysiphe graminis hordei*. *Crop Science*, **23**, 1115–19.
Mundt, C. C. & Browning, J. A. (1985). Development of crown rust epidemics in genetically diverse oat populations: effect of genotype unit area. *Phytopathology*, **75**, 607–10.
Mundt, C. C. & Leonard, K. J. (1985). Effect of host genotype unit area on epidemic development of crown rust following focal and general inoculations of mixtures of immune and susceptible oat plants. *Phytopathology*, **75**, 1141–5.
Muona, O., Allard, R. W. & Webster, R. K. (1982). Evolution of resistance to *Rhynchosporium secalis* (Oud.) Davis in barley Composite Cross II. *Theoretical and Applied Genetics*, **61**, 209–14.
Murphy, J. P., Helsel, D. B., Elliott, A., Thro, A. M. & Frey, K. J. (1982). Compositional stability of an oat multiline. *Euphytica*, **31**, 33–40.
Neergaard, P. (1977). *Seed Pathology*, vol. 1. London: MacMillan.
Nelson, R. R. (1973). The use of resistance genes to curb population shifts in plant pathogens. In *Breeding Plants for Disease Resistance*, ed. R. R. Nelson, pp. 49–66. University Park: Pennsylvania State University Press.
Nelson, R. R. (1978). Genetics of horizontal resistance to plant diseases. *Annual Review of Phytopathology*, **16**, 359–78.
Nelson, T. C. (1955). Chestnut replacement in the Southern Highlands. *Ecology*, **36**, 352–3.
Newhook, F. J. & Podger, F. D. (1972). The role of *Phytophthora cinnamomi* in Australian and New Zealand forests. *Annual Review of Phytopathology*, **16**, 299–326.
Niks, R. E. (1982). Early abortion of colonies of leaf rust, *Puccinia hordei*, in partially resistant barley seedlings. *Canadian Journal of Botany*, **60**, 714–23.
Noronha-Wagner, M. & Bettencourt, A. J. (1967). Genetic study of the resistance of *Coffea* spp. to leaf rust. I. Identification and behavior of four factors conditioning disease reaction in *Coffea arabica* to twelve physiologic races of *Hemileia vastatrix*. *Canadian Journal of Botany*, **45**, 2021–31.

Oates, J. D., Burdon, J. J. & Brouwer, J. B. (1983). Interactions between *Avena* and *Puccinia* species. II. The pathogens: *Puccinia coronata* Cda and *P. graminis* f. sp. *avenae* Eriks. and Henn. *Journal of Applied Ecology*, **20**, 585–96.
Ogle, H. J. & Brown, J. F. (1970). Relative ability of two strains of *Puccinia graminis tritici* to survive when mixed. *Annals of Applied Biology*, **66**, 273–9.
Ogle, H. J. & Brown, J. F. (1971). Some factors affecting the relative ability of two strains of *Puccinia graminis tritici* to survive when mixed. *Annals of Applied Biology*, **67**, 157–68.
Ohm, H. W. & Shaner, G. E. (1976). Three components of slow leaf-rusting at different growth stages in wheat. *Phytopathology*, **66**, 1356–60.
Oort, A. J. P. (1963). A gene-for-gene relationship in the *Triticum–Ustilago* system, and some remarks on host–parasite combinations in general. *Netherlands Journal of Plant Pathology*, **69**, 104–9.
Palti, J. (1981). *Cultural Practices and Infectious Crop Diseases*. Berlin: Springer-Verlag.
Palti, J. & Netzer, D. (1963). Development and control of *Phytophthora infestans* (Mont.) De By. under semi-arid conditions in Israel. *Phytopathology Mediterranea*, **2**, 265–74.
Pandey, M. C. & Wilcoxson, R. D. (1970). The effect of light and physiologic races on *Leptosphaerulina* leaf spot of alfalfa and selection for resistance. *Phytopathology*, **60**, 1456–62.
Parker, M. A. (1985). Local population differentiation for compatibility in an annual legume and its host-specific fungal pathogen. *Evolution*, **39**, 713–23.
Parker, M. A. (1986). Individual variation in pathogen attack and differential reproductive success in the annual legume, *Amphicarpaea bracteata*. *Oecologia* (*Berlin*), **69**, 253–9.
Parlevliét, J. E. (1977). Variation for partial resistance in a cultivar of rye, *Secale cereale*, to brown rust, *Puccinia recondita* f. sp. *recondita*. *Cereal Rusts Bulletin*, **5**, 13–16.
Parlevliet, J. E. (1979). Components of resistance that reduce the rate of epidemic development. *Annual Review of Phytopathology*, **17**, 203–22.
Parlevliet, J. E. (1981). Stabilizing selection in crop pathosystems: an empty concept or a reality? *Euphytica*, **30**, 259–69.
Parlevliet, J. E. & van Ommeron, A. (1975). Partial resistance of barley to leaf rust, *Puccinia hordei*. II. Relationship between field trials, micro plot tests and latent period. *Euphytica*, **24**, 293–303.
Pataky, J. M. & Lim, S. M. (1981). Effects of row width and plant growth habit on *Septoria* brown spot development and soybean yield. *Phytopathology*, **71**, 1051–6.
Paul, N. D. & Ayres, P. G. (1984). Effects of rust and post-infection drought on photosynthesis, growth and water relations in groundsel. *Plant Pathology*, **33**, 561–9.
Paul, N. D. & Ayres, P. G. (in press). The impact of a pathogen (*Puccinia lagenophorae*) on populations of groundsel (*Senecio vulgaris*) overwintering in the field. *Journal of Ecology*.
Pegg, G. F. (1981). The involvement of growth regulators in the diseased plant. In *Effects of Diseases on the Physiology of the Growing Plant*, ed. P. G. Ayres, pp. 149–78. Society for Experimental Biology Seminar Series, No. 11. Cambridge: Cambridge University Press.
Pelham, J. (1966). Resistance in tomato to tobacco mosaic virus. *Euphytica*, **15**, 258–67.
Person, C. (1959). Gene-for-gene relationships in host: parasite systems. *Canadian Journal of Botany*, **37**, 1101–30.
Person, C. (1966). Genetic polymorphism in parasitic systems. *Nature*, **212**, 266–7.
Person, C., Groth, J. V. & Mylyk, O. M. (1976). Genetic changes in host–parasite populations. *Annual Review of Phytopathology*, **14**, 177–88.
Person, C. & Mayo, G. M. E. (1974). Genetic limitations on models of specific interactions between a host and its parasite. *Canadian Journal of Botany*, **52**, 1339–47.

Person, C. & Sidhu, G. [S.] (1971). Genetics of host–parasite interrelationships. In *Mutation Breeding for Disease Resistance*, pp. 31–8. Vienna: International Atomic Energy Agency.

Pimentel, D. (1961). Animal population regulation by the genetic feed-back mechanism. *American Naturalist*, **95**, 65–79.

Podger, F. D. (1972). *Phytophthora cinnamomi*, a cause of lethal disease in indigenous plant communities in Western Australia. *Phytopathology*, **62**, 972–81.

Politowski, K. & Browning, J. A. (1978). Tolerance and resistance to plant disease: an epidemiological study. *Phytopathology*, **68**, 1177–85.

Populer, C. (1978). Changes in host susceptibility with time. In *Plant Disease: An Advanced Treatise*, vol. 2, ed. J. G. Horsfall & E. B. Cowling, pp. 239–62. New York: Academic Press.

Potter, L. R. (1980). The effects of barley yellow dwarf virus and powdery mildew in oats and barley with single and dual infections. *Annals of Applied Biology*, **94**, 11–17.

Potter, L. R. (1982). Interactions between barley yellow dwarf virus and rust in wheat, barley and oats, and the effects on grain yield and quality. *Annals of Applied Biology*, **100**, 321–9.

Powers, H. R., Jr & Sando, W. J. (1960). Genetic control of the host–parasite relationship in wheat powdery mildew. *Phytopathology*, **50**, 454–7.

Pratt, B. H., Heather, W. A. & Shepherd, C. J. (1973). Recovery of *Phytophthora cinnamomi* from native vegetation in a remote area of New South Wales. *Transactions of the British Mycological Society*, **60**, 197–204.

Price, P. W. (1980). *Evolutionary Biology of Parasites*. Princeton, New Jersey: Princeton University Press.

Ramakrishnan, L. (1966). Studies in the host–parasite relations of blast disease of rice. III. The effect of night temperature on the infection phase of blast disease. *Phytopathologische Zeitschrift*, **57**, 17–23.

Read, D. J. (1968). Some aspects of the relationship between shade and fungal pathogenicity in an epidemic disease of pines. *New Phytologist*, **67**, 39–48.

Regehr, D. L. & Bazzaz, F. A. (1979). The population dynamics of *Erigeron canadensis*, a successional winter annual. *Journal of Ecology*, **67**, 923–33.

Reichert, I. & Palti, J. (1967). Prediction of plant disease occurrence: a patho-geographical approach. *Mycopathologia*, **32**, 337–55.

Rice, W. R. (1983). Parent–offspring pathogen transmission: a selective agent promoting sexual reproduction. *American Naturalist*, **121**, 187–203.

Risser, G., Banihashemi, Z. & Davis, D. W. (1976). A proposed nomenclature of *Fusarium oxysporum* f. sp. *melonis* races and resistance genes in *Cucumis melo*. *Phytopathology*, **66**, 1105–6.

Robinson, R. A. (1976). *Plant Pathosystems*. Berlin: Springer-Verlag.

Roff, D. A. (1974). Spatial heterogeneity and the persistence of populations. *Oecologia (Berlin)*, **15**, 245–58.

Rosenthal, G. A. & Janzen, D. H. (ed.) (1979). *Herbivores: Their Interaction with Secondary Plant Metabolites*. New York: Academic Press.

Ross, J. P. (1983). Effect of soybean mosaic on component yields from blends of mosaic resistant and susceptible soybeans. *Crop Science*, **23**, 343–6.

Ross, M. A. & Harper, J. L. (1972). Occupation of biological space during seedling development. *Journal of Ecology*, **60**, 77–88.

Roth, E. R., Toole, E. R. & Hepting, G. H. (1948). Nutritional aspects of the little-leaf disease of pine. *Journal of Forestry*, **46**, 578–87.

Rovira, A. D. (1965). Plant root exudates and their influence upon soil microorganisms. In *Ecology of Soil-Borne Plant Pathogens*, ed. K. F. Baker & W. C. Synder, pp. 170–86. Berkeley: University of California Press.

Rowell, J. B. (1984). Controlled infection by *Puccinia graminis* f. sp. *tritici* under artificial conditions. In *The Cereal Rusts*, vol. 1, ed. W. R. Bushnell & A. P. Roelfs, pp. 291–332. New York: Academic Press.

Russell, G. E. (1978). *Plant Breeding for Pest and Disease Resistance*. London: Butterworths.

Sackston, W. E. (1962). Studies on sunflower rust. III. Occurrence, distribution and significance of races of *Puccinia helianthi* Schw. *Canadian Journal of Botany*, **40**, 1449–58.

Saghai Maroof, M. A., Webster, R. K. & Allard, R. W. (1983). Evolution of resistance to scald, powdery mildew, and net blotch in barley composite II populations. *Theoretical and Applied Genetics*, **66**, 279–83.

Samborski, D. J. & Dyck, P. L. (1968). Inheritance of virulence in wheat leaf rust on the standard differential wheat varieties. *Canadian Journal of Genetics and Cytology*, **10**, 24–32.

Samborski, D. J. & Ostapyk, W. (1959). Expression of leaf rust resistance in Selkirk and Exchange wheats at different stages of plant development. *Canadian Journal of Botany*, **37**, 1153–5.

Sandfaer, J. (1968). Induced sterility as a factor in the competition between barley varieties. *Nature*, **218**, 241–3.

Sandfaer, J. (1970). Barley stripe mosaic virus as the cause of sterility interaction between barley varieties. *Hereditas*, **64**, 150–2.

Sarukhan, J. & Harper, J. L. (1973). Studies on plant demography: *Ranunculus repens* L., *R. bulbosus* L. and *R. acris* L. I. Population flux and survivorship. *Journal of Ecology*, **61**, 675–716.

Schafer, J. F. (1971). Tolerance to plant disease. *Annual Review of Phytopathology*, **9**, 235–52.

Schmidt, R. A. (1978). Diseases in forest ecosystems: the importance of functional diversity. In *Plant Disease: An Advanced Treatise*, vol. 2, ed. J. G. Horsfall & E. B. Cowling, pp. 287–315. New York: Academic Press.

Schoeneweiss, D. F. (1975). Predisposition, stress, and plant disease. *Annual Review of Phytopathology*, **13**, 193–211.

Scott, P. R., Johnson, R., Wolfe, M. S., Lowe, H. J. B. & Bennett, F. G. A. (1979). Host-specificity in cereal parasites in relation to their control. *Plant Breeding Institute, Cambridge, Annual Report*, 1978, 27–62.

Scott, S. W. & Griffiths, E. (1980). Effects of controlled epidemics of powdery mildew on grain yield of spring barley. *Annals of Applied Biology*, **94**, 19–31.

Sedcole, J. R. (1978). Selection pressures and plant pathogens: stability of equilibria. *Phytopathology*, **68**, 967–70.

Shaner, G. (1973). Reduced infectability and inoculum production as factors of slow mildewing in Knox wheat. *Phytopathology*, **63**, 1307–11.

Sharitz, R. R. & McCormick, J. F. (1973). Population dynamics of two competing annual plant species. *Ecology*, **54**, 723–40.

Sharma, J. K., Heather, W. A. & Winer, P. (1980). Effect of leaf maturity and shoot age of clones of *Populus* species on susceptibility to *Melampsora larici-populina*. *Phytopathology*, **70**, 548–54.

Shattock, R. C. (1974). Variation and its origins in *Phytophthora infestans* (Mont.) de Bary. Ph.D. thesis, University of Wales.

Shepherd, R. J. (1972). Transmission of viruses through seed and pollen. In *Principles and Techniques of Plant Virology*, ed. C. I. Kado & H. O. Agrawal, pp. 267–92. Princeton, New Jersey: Van Nostrand-Reinhold.

Shukla, V. D. & Anjaneyulu, A. (1981). Plant spacing to reduce rice tungro incidence. *Plant Disease*, **65**, 584–6.

Siddiqui, M. Q. & Manners, J. G. (1971). Some effects of general yellow rust (*Puccinia striiformis*) infection on ^{14}carbon assimilation, translocation and growth in a spring wheat. *Journal of Experimental Botany*, **22**, 792–9.

Sidhu, G. S. (1975). Gene-for-gene relationships in plant parasitic systems. *Science Progress (Oxford)*, **62**, 467–85.

Sidhu, G. S. (1980). Genetic analysis of plant parasitic systems. *XIV International Congress of Genetics*, Moscow, 1978, vol. 1, pp. 391–408. Moscow: MIR Publishers.

Sidhu, G. S. (1981). Genetic resistance and disease complexes. In *Proceedings of Symposia, Ninth International Congress of Plant Protection, Washington, D.C., USA*, August 1979, ed. T. Kommedahl, pp. 182–5. College Park, Maryland: Entomology Society of America.

Sidhu, G. [S.] & Person, C. (1972). Genetic control of virulence in *Ustilago hordei*. III. Identification of genes for host resistance and demonstration of gene-for-gene relations. *Canadian Journal of Genetics and Cytology*, **14**, 209–13.

Sidhu, G. S. & Webster, J. M. (1977). Genetics of simple and complex host–parasite interactions. In *Induced Mutations Against Plant Diseases*, pp. 59–79. Vienna: International Atomic Energy Agency.

Silvertown, J. W. (1982). *Introduction to Plant Population Ecology*. London: Longmans.

Simons, M. D. (1970). *Crown Rust of Oats and Grasses*. Monograph No. 5, American Phytopathology Society. Worcester, Massachusetts: American Phytopathological Society.

Smith, D. M. (1951). *The Influence of Seedbed Conditions on the Regeneration of Eastern White Pine*. Bulletin no. 545, Connecticut Agricultural Station.

Smith, H. C. & Blair, I. D. (1950). Wheat powdery mildew investigations. *Annals of Applied Biology*, **37**, 570–83.

Snaydon, R. W. & Davies, M. S. (1972). Rapid population differentiation in a mosaic environment. II. Morphological variation in *Anthoxanthum odoratum*. *Evolution*, **26**, 390–405.

Snaydon, R. W. & Davies, M. S. (1976). Rapid population differentiation in a mosaic environment. IV. Populations of *Anthoxanthum odoratum* at sharp boundaries. *Heredity*, **37**, 9–25.

Snow, J. A. (1964). Effects of light on initiation and development of bean rust disease. *Dissertation Abstracts*, **25**, 6149.

Stakman, E. C. & Harrar, J. G. (1957). *Principles of Plant Pathology*. New York: Ronald Press.

Steadman, J. R., Coyne, D. P. & Cook, G. E. (1973). Reduction of severity of white mold disease on great northern beans by wider row spacing and determinate plant growth habit. *Plant Disease Reporter*, **57**, 1070–1.

Strong, D. R., Jr & Levin, D. A. (1979). Species richness of plant parasites and growth form of their hosts. *American Naturalist*, **114**, 1–22.

Suneson, C.A. (1969). Registration of barley composite crosses. *Crop Science*, **9**, 395–6.

Suneson, C. A. & Wiebe, G. A. (1942). Survival of barley and wheat varieties in mixtures. *Agronomy Journal*, **34**, 1052–6.

Tansley, A. G. & Adamson, R. S. (1925). Studies of the vegetation of the English chalk. III. The chalk grasslands of the Hampshire–Sussex border. *Journal of Ecology*, **13**, 177–223.

Tenne, F. D., Prasartsee, C., Machado, C. C. & Sinclair, J. B. (1974). Variation in germination and seedborne pathogens among soybean seed lots from three regions in Illinois. *Plant Disease Reporter*, **58**, 411–13.

Thresh, J. M. (1976). Gradients of plant virus diseases. *Annals of Applied Biology*, **82**, 381–406.

Timmer, L. W. & Fucik, J. E. (1975). The effect of rainfall, drainage, tree spacing and fungicide application on the incidence of citrus brown rot. *Phytopathology*, **65**, 241–2.

Trenbath, B. R. (1974). Biomass productivity in mixtures. *Advances in Agronomy*, **26**, 177–210.

van den Bergh, J. P. & Ennik, G. C. (1973). Vzaimootnoshenie mezdhu vidami rastenii. In *Fiziologo-biokhimicheskie osnovy vzaimodeistviya rastenii v fitotsenozakh. Vipusk*, **4**, 47. (Unpublished English translation supplied by senior author.)

Vandermeer, J. H. (1973). On the regional stabilization of locally unstable predator–prey relationships. *Journal of Theoretical Biology*, **41**, 161–70.

van der Plank, J. E. (1948). The relation between the size of fields and the spread of plant-disease into them. Part I. Crowd diseases. *Empire Journal of Experimental Agriculture*, **16**, 134–42.

van der Plank, J. E. (1949). The relation between the size of fields and the spread of plant-diseases into them. Part II. Diseases caused by fungi with air-borne spores; with a note on horizons of infection. *Empire Journal of Experimental Agriculture*, **17**, 18–22.

van der Plank, J. E. (1960). Analysis of epidemics. In *Plant Pathology: An Advanced Treatise*, vol. 3, ed. J. G. Horsfall & A. E. Dimond, pp. 229–89. New York: Academic Press.

van der Plank, J. E. (1963). *Plant diseases: Epidemics and Control*. New York: Academic Press.

van der Plank, J. E. (1968). *Disease Resistance in Plants*. New York: Academic Press.

van der Plank, J. E. (1978). *Genetic and Molecular Basis of Plant Pathogenesis*. Berlin: Springer-Verlag.

Waggoner, P. E. (1962). Weather, space, time, and chance of infection. *Phytopathology*, **52**, 1100–8.

Wahl, I. (1970). Prevalence and geographic distribution of resistance to crown rust in *Avena sterilis*. *Phytopathology*, **60**, 746–9.

Wahl, I., Eshed, N., Segal, A. & Sobel, Z. (1978). Significance of wild relatives of small grains and other wild grasses in cereal powdery mildews. In *The Powdery Mildews*, ed. D. M. Spencer, pp. 83–100. London: Academic Press.

Walker, J. C. & Wellman, F. L. (1926). Relation of temperature to spore germination and growth of *Urocystis cepulae*. *Journal of Agricultural Research, Washington*, **32**, 133–46.

Waloff, N. & Richards, O. W. (1977). The effect of insect fauna on growth, mortality and natality of broom, *Sarothamnus scoparius*. *Journal of Applied Ecology*, **14**, 787–98.

Walters, D. R. (1985). Shoot:root interrelationships: the effects of obligately biotrophic fungal pathogens. *Biological Reviews*, **60**, 47–79.

Walters, D. R. & Ayres, P. G. (1981). Growth and branching pattern of roots of barley infected with powdery mildew. *Annals of Botany N.S.*, **47**, 159–62.

Watson, D. J. & Wilson, J. H. (1956). An analysis of the effects of infection with leaf-roll virus on the growth and yield of potato plants, and of its interactions with nutrient supply and shading. *Annals of Applied Biology*, **44**, 390–409.

Watson, I. A. & Singh, D. (1952). The future for rust resistant wheat in Australia. *Journal of the Australian Institute of Agricultural Science*, **18**, 190–7.

Weaver, S. E. & Warwick, S. I. (1982). Competitive relationships between atrazine resistant and susceptible populations of *Amaranthus retroflexus* and *A. powellii* from southern Ontario. *New Phytologist*, **92**, 131–9.

Webster, R. K., Hall, D. H., Heeres, J., Wick, C. M. & Brandon, D. M. (1970). *Achlya kelbsiana* and *Pythium* species as primary causes of seed rot and seedling disease of rice in California. *Phytopathology*, **60**, 964–8.

Weir, B. S., Allard, R. W. & Kahler, A. L. (1972). Analysis of complex allozyme polymorphisms in a barley population. *Genetics*, **72**, 505–23.

Weir, B. S., Allard, R. W. & Kahler, A. L. (1974). Further analysis of complex polymorphisms in a barley population. *Genetics*, **78**, 911–19.

Wells, O. O., Switzer, G. L. & Nance, W. L. (1982). Genotype–environment interaction in rust resistance in Mississippi loblolly pine. *Forest Science*, **28**, 797–809.

Weltzien, H. C. (1972). Geophytopathology. *Annual Review of Phytopathology*, **10**, 277–98.

Weste, G. (1974). *Phytophthora cinnamomi* – the cause of severe disease in certain native communities in Victoria. *Australian Journal of Botany*, **22**, 1–8.

Weste, G. (1980). Vegetation changes as a result of invasion of forest on Krasnozem by *Phytophthora cinnamomi*. *Australian Journal of Botany*, **28**, 139–50.

Weste, G. (1981). Changes in the vegetation of sclerophyll shrubby woodland associated with invasion by *Phytophthora cinnamomi*. *Australian Journal of Botany*, **29**, 261–76.

Weste, G., Cooke, D. & Taylor, P. (1973). The invasion of native forest by *Phytophthora cinnamomi*. II. Post-infection vegetation patterns, regeneration, decline in inoculum and attempted control. *Australian Journal of Botany*, **21**, 13–29.

White, E. M. (1982). The effects of mixing barley cultivars on incidence of powdery mildew (*Erysiphe graminis*) and on yield in Northern Ireland. *Annals of Applied Biology*, **101**, 539–45.

White, J. (1980). Demographic factors in populations of plants. In *Demography and Evolution in Plant Populations*, ed. O. T. Solbrig, pp. 21–48. California: University of California Press.

White, J. & Harper, J. L. (1970). Correlated changes in plant size and number in plant populations. *Journal of Ecology*, **58**, 467–85.

Whittaker, J. B. (1982). The effect of grazing by a chrysomelid beetle, *Gastrophysa viridula*, on growth and survival of *Rumex crispus* on a shingle bank. *Journal of Ecology*, **70**, 291–6.

Williams, G. C. (1975). *Sex and Evolution*. Princeton, New Jersey: Princeton University Press.

Wolfe, M. S. & Barrett, J. A. (1980). Can we lead the pathogen astray? *Plant Disease*, **64**, 148–55.

Wolfe, M. S., Barrett, J. A. & Jenkins, J. E. E. (1981). The use of cultivar mixtures for disease control. In *Strategies for the Control of Cereal Disease*, ed. J. F. Jenkyn & R. T. Plumb, pp. 73–80. Oxford: Blackwell Scientific Publications.

Wolfe, M. S. & Knott, D. R. (1982). Populations of plant pathogens: some constraints on analysis of variation in pathogenicity. *Plant Pathology*, **31**, 79–90.

Wolfe, M. S. & Minchin, P. N. (1976). Quantitative assessment of variation in field populations of *Erysiphe graminis* f. sp. *hordei* using mobile nurseries. *Transactions of the British Mycological Society*, **66**, 332–4.

Woods, F. W. & Shanks, R. E. (1957). Replacement of chestnut in the Great Smoky Mountains of Tennessee and North Virginia. *Journal of Forestry*, **55**, 11.

Yarwood, C. E. (1956). Heat-induced susceptibility of beans to some viruses and fungi. *Phytopathology*, **46**, 523–5.

Yarwood, C. E. (1959). Predisposition. In *Plant Pathology: An Advanced Treatise*, vol. 1, ed. J. G. Horsfall & A. E. Dimond, pp. 521–62. New York: Academic Press.

Yoda, K., Kira, T., Ogawa, H. & Hozumi, K. (1963). Self-thinning in overcrowded pure stands under cultivated and natural conditions. *Journal of Biology, Osaka City University*, **14**, 107–29.

Young, H. C., Jr & Prescott, J. M. (1977). A study of race populations of *Puccinia recondita* f. sp. *tritici*. *Phytopathology*, **67**, 528–32.

Zadoks, J. C. (1961). Yellow rust on wheat: studies in epidemiology and physiological specialization. *Netherlands Journal of Plant Pathology*, **67**, 69–256.

Zadoks, J. C. & Kampmeijer, P. (1977). The role of crop populations and their development, illustrated by means of a simulator, Epimul 76. *Annals of the New York Academy of Science*, **287**, 164–90.
Zadoks, J. C. & Schein, R. D. (1979). *Epidemiology and Plant Disease Management*. New York: Oxford University Press.
Zentmyer, G. A. (1961). Chemotaxis of zoospores for root exudates. *Science*, **133**, 1595–6.
Zimmer, D. E. & Rehder, D. (1976). Rust resistance of wild *Helianthus* species of the north central United States. *Phytopathology*, **66**, 208–11.

Index

Page numbers in bold type refer to entries in the glossary.